MUSTANG
RESTORATION
HANDBOOK

by Don Taylor & Tom Wilson

HPBooks

HPBooks
Published by the Penguin Group
Penguin Group (USA) Inc.
375 Hudson Street, New York, New York, 10014, USA
Penguin Group (Canada), 90 Eglinton Avenue East, Suite 700, Toronto, Ontario M4P 2Y3, Canada
(a division of Pearson Penguin Canada Inc.)
Penguin Books Ltd., 80 Strand, London WC2R 0RL, England
Penguin Group Ireland, 25 St. Stephen's Green, Dublin 2, Ireland (a division of Penguin Books Ltd.)
Penguin Group (Australia), 250 Camberwell Road, Camberwell, Victoria 3124, Australia
(a division of Pearson Australia Group Pty. Ltd.)
Penguin Books India Pvt. Ltd., 11 Community Centre, Panchsheel Park, New Delhi—110 017, India
Penguin Group (NZ), 67 Apollo Drive, Mairangi Bay, Auckland 1311, New Zealand
(a division of Pearson New Zealand Ltd.)
Penguin Books (South Africa) (Pty.) Ltd., 24 Sturdee Avenue, Rosebank, Johannesburg 2196,
South Africa

Penguin Books Ltd., Registered Offices: 80 Strand, London WC2R 0RL, England

While the author has made every effort to provide accurate telephone numbers and Internet addresses at the time of publication, neither the publisher nor the author assumes any responsibility for errors, or for changes that occur after publication. Further, publisher does not have any control over and does not assume any responsibility for author or third-party websites or their content.

First edition: January 1987

Library of Congress Cataloging-in-Publication Data

Taylor, Don (Donald D.)
  Mustang restoration handbook.
  At head of title: HPBooks
  Includes index.
  1. Mustang automobile.   Automobiles—Conservation
and restoration.   I. Wilson, Tom (Tom S.)   II. HPBook (Firm)
III. Title.   IV. Title: HPBooks Mustang restoration handbook.
TL215.M8T39   1987   629.28'722   87-195
ISBN 978-0-89586-402-4

PRINTED IN THE UNITED STATES OF AMERICA

40   39

NOTICE: The information in this book is true and complete to the best of our knowledge. All recommendations on parts and procedures are made without any guarantees on the part of the author or the publisher. Tampering with, altering, modifying or removing any emissions-control device is a violation of federal law. Author and publisher disclaim all liability incurred in connection with the use of this information.

Contents

CONTRIBUTORS LIST

To make a book like this possible requires the help of many professionals. Below is a list of those professionals, and the companies they represent, who contributed directly through time, talent and product to the success of this book. I thank you all and owe each of you a debt of gratitude.

Cloyes Gear
Mike Tarascot

Rinshed Mason (R/M Paints)
Al Rivera

Sealed Power
Gary Wade

Federal Mogul
Dan Ross

SKF Automotive
Walt Delevich

Fel-Pro, Inc.
Robert Morris

Acme Headliner
George Westmorland

California Mustang
Steve Bennett

Mustang Classics
Dave & Arlene Cawthorne

Capital Auto Repair
Duane & Jim Hammes

Carr's Auto Parts
Carl Carr & Tim Mints

Photo Darkroom
Bob Hill

Gauthier Automotive
Dave Gauthier

Hughes Paint & Tool
Dick Hughes

San Diego Mustang Club
Bill Smith

Hilltop Classics
Jerry Olmsted

Mustangs: 1965-1970
Dave Stroot

Mustang Car Club (MC[2])
Ken Wittick

East Mission Auto Body
Bruce Chamness
Timothy O'Rourke

Sun Shield of California
Ed Cohan

Goodyear Tire & Rubber
Bob Sealy

American Racing Equipment
Bill Madden

Mustangs Unlimited
John Browning

Tucson Frame Service
Don Kott

Mustang Majic
Jim Brunk

Arizona Mustang Supply
Don Laufenburger

Council Street Automotive
Marv Kea

Southern Arizona Mustang Club
Dave Carroll
Dave Hoverstock
Larry Wren

Jim Osborn Reproductions
Jim Osborn

The Eastwood Company
Curt Strohaker

Stainless Steel Brakes Corp.
Lauren H. Jonas

Dallas Mustang
Dick Meditz

And a special thanks to Lewis Stilwell, Jim Beebe and Bob Lyndal

History & Identification

The first Mustangs struck a nerve in this country and created a whole, new market for small, sporty, performance cars—*ponycars.* Here are two of the more desirable early cars—a '66 GT 2+2 fastback and a '65 K-model convertible. Photo by Ron Sessions.

On April 17, 1964, after one of the grandest media blitzes ever, Ford introduced the newest automotive concept to the American public: the original *pony car.* It was the 1964-1/2 Mustang—officially introduced as a '65 model. This sporty offshoot of the diminutive Ford Falcon captured the buying public's imagination in an even greater way than did its two-seat Thunderbird predecessor. For less than $2500 you could have a car with more than 35 *standard* items. This, coupled with a list of 40 options, made the Mustang easily tailored to individual tastes.

Ford introduced its new sporty car in coupe and convertible form. It wasn't until five months later in September that the fastback (2+2) was added. Specialty models would come and go, but the coupe, convertible and fastback remained throughout those exciting years of the late '60s and early '70s.

The wheelbase of the new Mustang was only 6-in. longer than that of the '57 Thunderbird, yet it could carry four passengers in comfort—as long as the rear passengers were relatively small. With its long hood line, upswept rear end and lovely sculptured sides, it became the only automobile (or commercial product ever) to receive the Tiffany Award for Excellence in American Design. It is this excellence of design, simplicity of function and availability that makes the early Mustang such a hobbyist's dream. Let's look at some of the things that made the car what it is.

1965

The Mustang's 35 standard features included such items as: front bucket seats, floor shift (automatic or manual, bucket or optional bench seats), instrument-panel dash pad, molded carpets, vinyl upholstery, choice of two color-keyed interiors, color-keyed seat belts, full wheel covers and self-adjusting brakes. A choice of six different engine configurations and seventeen exterior colors gave prospective buyers just about all they could ask for.

Body Styling—The evolution of the Mustang began with race and sports-car prototypes. Its heritage was born of the Mustang I, a 90-in. wheelbase, two-passenger sports car with a rear-mounted engine. A large functional air scoop on each rear

Although the first Mustang was technically a 1964-1/2, 1965 was the first full production year. And Ford registered them all as '65s.

The fastback 2 + 2 caught Carroll Shelby's imagination right off. He, of course, went on to produce the fabulously successful Shelby Mustangs.

fender directed air through the rear-mounted engine's radiator. This was followed in 1963 with the Falcon based front-engine Mustang II, looking almost the way the production 1965 model would look.

The functional air scoop, however, gave way to a sculptured side panel when the engine was moved to the front on the Mustang II. The wheelbase was extended to 108 in., to make room for a rear seat. This car, like its 1957 Thunderbird cousin, had a removable hardtop. Unlike the production model, however, it had no bumpers! Within 18 months, these two prototypes were history and the production models began to roll out of the factory.

Styling of the three production models allowed for some interchangeablility between cars. The windshields, front end and doors on all three models were common. However, each body style incorporated a different door window and rear quarter window. Keep this in mind if you must replace a door.

The fastback did not have a rear quarter window as such. Instead, it had *air extractors*—louvers that allowed for air circulation in the rear compartment of the car. For "air extraction" in the convertible you lowered the top or unzipped the rear window. The top was manually operated with a power top optional.

Interior—What were expensive interior options in other Detroit cars came standard in the '65 Mustang. This included such appointments as bucket seats (a bench seat with fold-down armrest was an option), molded carpets, floor-mounted shifter

(even with the bench seat), instrument-panel dash pad and a multitude of small items such as armrests, sunvisors, glove-compartment light and sports steering wheel.

You could have your choice of five all-vinyl interiors. These included red, white, blue, black or a buckskin color called *palomino*. If you preferred cloth inserts, these came in black or palomino. Later, palomino would be dropped in favor of a color called *parchment*; and a sixth color added, ivy gold.

In March, 1965, a whole new interior was introduced. Ford listed it as the *Decorator Interior Group*, but enthusiasts dubbed it the *pony interior* from its richly embossed logo in the center of the seatback—a running herd of ponies.

Instrumentation of the new Mustang was clustered above the steering wheel with the central component a 120-mph speedometer. Detroit was still using gages for fuel and temperature but had succumbed to the use of warning lights for oil pressure and charging-system amperage. If you had a '65 model built in the first part of 1964, the light would read GEN as those models had not yet been changed from a d-c-generator charging system to an a-c-generator—alternator—charging system. The alternator warning light read ALT.

To further enhance the instrumentation, an optional electric tachometer and 24-hour clock could be ordered. Called the *Rally-Pac*, it was mounted across the steering column. Standard calibration for the tachometer was 6000 rpm. If you opted for the

A bench seat in a Mustang? That's right. Though never popular, Ford offered one in its pony car for several years. Here's a '66 with the desirable Pony Interior. Photo by Ron Sessions.

289-CID High Performance engine, the tachometer was increased to an 8000-rpm unit. Later, a low-profile Rally-Pac would become available as part of the GT Equipment Group option.

Drive Trains—The Mustang was designed to be powered and sold as a six-cylinder automobile—like the Falcon it

Showing its Ford Falcon heritage, many early Mustangs were six-bangers, like this '66 200-CID T-5. See page 19. Photo by Ron Sessions.

Probably the most popular Mustang engine of all time, the Ford 289-CID V8. Aftermarket Monte Carlo bar running between fenders is a non-stock item.

Standard equipment on the '65—67 G.T.350 was a 306-HP, 289-CID small-block V8 with a Holley 4150 carburetor, high-rise aluminum intake manifold, headers, aluminum valve covers and more. Other Shelby upgrades shown on Dave Haverstocks' unrestored '66 G.T.350 are a Monte Carlo bar and export brace. Photo by Ron Sessions.

was based on. In practice, this was not the case. More than 70% of the buyers chose one of three V8s. The engine lineup for 1965 included 170- and 200-CID six-cylinder engines. These were teamed with a standard three-speed manual transmission or an optional British-built Dagenham four-speed manual or Cruise-O-Matic.

If you elected instead for one of the V8s, you had a choice between a 260-CID two-barrel, 289-CID two-barrel, 289-CID four-barrel, and the High Performance (*HiPo*) 289. As standard equipment, the 260, 289 two-barrel, and 289 four-barrel engines came with the three-speed manual transmission. The four-speed—a U.S. built top-loader—was optional with only the two 289s and HiPo 289. The Cruise-O-Matic was optional on all but the HiPo 289.

Later in 1965, the 260-CID V8 and 170-CID six were dropped.

GT Equipment Group—In celebration of the Mustang's first birthday, Ford introduced two new options. The first was mentioned earlier as the *pony interior*. The second was the new performance and appearance group called the *GT option*, or *GT package*. All of the goodies in this package were already available as regular options. Ford, however, assembled them together on the car at the factory which made it a package rather than a random group of options.

What did this package include? Well, it began with the 289-CID four-barrel or 289 HiPo engine at 225 or 271 HP, respectively. This ensured top performance. To signal the fact that this was more than just your ordinary family car, the factory included: polished stainless-steel dual exhausts that exited through the rear valance panel, two grille-mounted fog lights and a "GT" stripe running between the front and rear wheel openings, just above the rocker panel.

Up front were four-piston, fixed-caliper disc brakes and a brake pedal with a center logo indicating the same. The instrument cluster was changed to a five-dial gage setup in black and chrome. If you ordered the pony interior, the cluster was trimmed in walnut.

Shelby G.T.350—Enter Carroll Shelby: race-car driver, race-car designer and founder of Shelby-American, Inc. He successfully married a British race-car design to an American powerplant, creating the world-famous Cobra. By shoehorning a Ford 260-CID V8 into an AC Bristol, Shel-

For 1966, 10-spoke aluminum wheels were available for G.T.350. Photo by Ron Sessions.

by began a threat to modern sports-car racing that would carry over into the Shelby G.T.350 of 1965.

The Shelby G.T.350 began life as a regular Mustang 2 + 2, shipped to Shelby-American minus the hood. There, an all-fiberglass, latchless replacement, with functional hood scoop and hood pins, was installed initially. Beneath this sat a modified High Performance 289 engine rated at 306 HP. Shelby added an aluminum high-rise intake manifold and a 715-cfm Holley 4150 carburetor with center-pivot float, and hood pins, a large-capacity, baffled cast-aluminum oil pan, exhaust headers and "glasspack" mufflers. The exhaust dumped out at each side just in front of the rear wheels. For appearance, the intake manifold, oil pan and aluminum valve covers were cast with the word COBRA across them.

An aluminum-case Borg-Warner T-10 with close-ratio gears (later known as the "Sebring" transmission) and Detroit Automotive "No Spin" ratcheting differential got the power to the pavement. And above-the-axle rear traction bars were added to help control rear wheel hop upon hard acceleration. Also added were rebound-limiting rear-axle cables and a drive-shaft strap.

Shelby added his own touch to the front end, improving the car's roadworthiness by moving the upper-control-arm inner pivot down 1 in. to give more negative camber in jounce and effectively lowering the front end about 3/4 in. He also replaced the stock 3/4-in. front sway bar with a stout 1-in. bar. Dutch-made, adjustable Koni shocks were added front and rear, as were Goodyear

7.75 x 15 Blue Dot tires, rated for 130 mph and mounted on 5-1/2-in. station-wagon wheels. Stopping power was provided by a set of 11.3-in. Kelsey-Hayes front discs with ventilated rotors and metallic pads, and larger passenger-car drums with segmented metallic drum-brake linings at the rear.

Reinforcing the front end and reducing deflections was a one-piece *export* brace (previously used only in Mustangs sold overseas) to triangulate the shock towers and firewall, as well as a *Monte Carlo* bar running between the front spring towers.

The all-vinyl interior was designed to be functional and was available in any color as long as it was black. A wood-rimmed Shelby-American aluminum steering wheel replaced the deep-dished Mustang unit. Standard Mustang instrumentation was supplanted by a large-dial tachometer and oil-pressure gage, housed in a large, add-on pod atop the instrument panel to the right of the driver. Seat belts were wide-web aircraft grade. On most cars, the rear seat was deleted and the spare tire was moved out of the trunk and placed on a fiberglass shelf in the rear-seat area. About 300 cars had trunk-mounted batteries to improve weight distribution.

Single color availability carried to the exterior, where wimbledon white with twin guardsman blue rocker-panel racing stripes was standard dress. The standard bright Mustang grille was removed, leaving the black anodized, honeycomb unit. A pair of wide "runway" stripes running longitudinally over the hood, roof and decklid, also in guardsman blue, were optional.

The Shelby G.T.350 should not be confused with the GT package that was available on all Mustangs. They were two separate products: one by Carroll Shelby, the other by Ford.

1966

1966 was a year of refinement with no major sheet-metal changes. Parts that were interchangeable in 1965 remained interchangeable in 1966; and, with their 1965 counterparts. The refinements then, were limited to minor trim items, or functional under-the-skin features. Basic design was left alone.

Body Styling—The blacked-out honeycomb grille of '65 was replaced with an extruded-aluminum grille. Except for the GT, each horizontal grille bar had exposed bright strips.

The large horizontal crossbar of the '65's was dropped as was the short vertical bar, but the running mustang remained. However, the '66 GT package, the familiar

Performance was king in Detroit's pre-Nader musclecar era. For '66, Ford made a 140-mph speedometer standard equipment in the Mustang. Photo by Ron Sessions.

crossbar between each light and mustang was used.

The side sculpture, which in 1965 was simply an indentation in the sheet metal, was now filled in on some models with a three-prong, chromed, simulated air scoop. This gave the appearance of a brake-cooling duct often used on racing machines.

Rocker-panel moldings were added in this year, giving the car a dressier look.

Interior—Interior changes in '66 were more extensive than those of the exterior. The instrument cluster was replaced with five-gage panel that eliminated the warning lights of the previous year. There were new gages for oil pressure and an ammeter. And a new 140-mph speedometer was standard fare—even on six-cylinder models!

The instrument-panel dash pad was reshaped, as was the glovebox and glovebox door. The optional low-profile Rally-Pac, introduced in 1965, remained an option in 1966, but without its identifying RALLY-PAC script and "eyebrows."

Changes in upholstery styling were minor. What had been vertical pleats in the door panels in 1965, became horizontal in 1966. The smooth vinyl seat inserts became knit-weave vinyl inserts in 1966. The pony interior remained as it was in '65. In 1965, the VIN plate numbered 28 trim schemes. This had increased to 35 in 1966.

GT Package—The GT equipment group remained much as it was in 1965. The '65 horizontal grille bar between the grille-mounted fog lights were retained for 1966. Again, two 289s were the powerplants used in the GTs, although they were not part of the option. The rest of the option package included: three- or four-speed transmis-

For '65, G.T.350s came but one way—Wimbleton White with Guardsman Blue stripes. But in '66, Shelbys could be had in a variety of colors, including this Raven Black example with gold stripes. Photo by Ron Sessions.

On '66 Shelby G.T.350s, side scoop for rear brake cooling was functional. Photo by Ron Sessions.

Shelby first spotted this quarter window conversion in a Ford Motor Company concept Mustang, and liked it so much, he put it into production for '66. Photo by Ron Sessions.

Spartan cockpit was mark of '65—66 G.T.350s, with black vinyl upholstery, aircraft-grade seat belts and dash-mounted tachometer as standard fare. Photo by Ron Sessions.

sion; front disc brakes; a special handling package; stainless-steel, flared tailpipe extensions terminating the low restriction dual mufflers; 4-in. fog lamps; GT stripes; the word MUSTANG spelled out on each fender; and a GT badge mounted directly above it. The package terminated with a special gas cap with the GT script replacing the Mustang insignia.

Drive Train—With the less popular 170-CID six-cylinder and 260-CID V8 dropped from the line in late-'65, this left the 200-CID six as standard with the 289-CID two-barrel and four-barrel, plus the HiPo 289 as options. The three-speed manual transmission was standard on all but the HiPo 289 engine. The four-speed manual and beefed-up C-4 Cruise-O-Matic were optional with all four engines.

Shelby—In 1966, Ford wanted Shelby to show a profit—or at least break even! To do this, some content had to go and a great-er audience had to be attracted. The racing aspects of the car were elected. Thus, the Detroit Locker rear end was put on the option list and the noisy side-exiting exhaust were deleted. Shelby also stopped lowering the front A-arm pivots in '66. And later that year, the over-axle rear traction bars were deleted in favor of more-conventional under-axle bars.

Exterior changes amounted to adding functional rear quarter-panel scoops for rear-brake cooling, replacing the louvers with clear plexiglass quarter windows and adding a new G.T.350 gas cap. The customer no longer had to endure the single choice of a white and blue paint job. He could now select from four new colors: raven black, guardsman blue, ivy green and candy apple red.

A succession of hood designs were tried. Because the all-fiberglass hood of the '65s tended to warp and crack, early '66 models were treated to a fiberglass hood molded over a steel inner panel. These were less than durable as well, so late '66s were equipped with *all-steel* hoods.

The basic change to the interior was to adopt the standard Mustang's five-dial instrument panel. This eliminated the tachometer/oil-pressure bezel. A special 9000-rpm tachometer was bolted to the top of the dash pad, and oil pressure was monitored by the existing gage in the new instrument cluster.

Shelby's Cobra engine remained the same as it was in '65. However, a few lucky buyers were able to get their new Shelby equipped with a Paxton supercharger! This belt-driven supercharger, feeding an Autolite 460-cfm carburetor in a special Cobra enclosure, added 46% more HP!

If you were very, very, lucky and a friend of Carroll Shelby, you might have received one of the six convertibles he made in 1966. Count them friends, six Shelby Convertibles!

G.T.350H—The G.T.350 did indeed show a profit: by striking a deal with Hertz Rent-A-Car for 936 units. Although a few other colors were produced, most of the cars were black. All had the gold G.T.350H side stripe, while most had the gold racing stripes. Chrome Magnum 500 wheels were used, as opposed to the dull-gray finish Magnums used on regular Shelbys. The majority of the Hertz cars were equipped with C-4 automatics and Autolite four-barrel carburetors. Some of the earliest models, though, had four-speed manual transmissions.

1967

Although distinctively Mustang, the 1967 model underwent a major facelift. In fact, so much so that *most* body parts wouldn't interchange with '65—66 models. Furthermore, 1967 was the first year the FE big-block engine—the 390 Thunderbird Special—was available in the Mustang. It was to be followed by even larger and more-powerful engines as the muscle-car era came of age.

Body Styling—Although Mustang still

Early Shelbys could be "accessorized" with numerous goodies from the Shelby-American catalog. Here, Larry Wren's prize-winning '66 G.T.350 sports four downdraft 2-barrel Webers. Photo by Ron Sessions.

On a warm, summer night, what can be more fun to drive than a Mustang convertible? Take this lovely 1967 example, for instance.

sported a 108-in. wheelbase, little else was carried over. For '67, there was new sheet metal, a new grille, new taillights, new side and rear glass, new bumpers and chrome trim. Perhaps the major body change was a 3-in. increase in overall width to about 80 in. This provided room for the larger 390-CID engine and wider tires.

Other significant changes took place in the fastback 2+2. Here, the fastback no longer terminated at the leading edge of the deck lid. Instead, it flowed smoothly to the rear edge of the deck lid. This was more in keeping with the traditional definition of a fastback. Below the deck lid, the rear panel was made concave.

Although not a body-styling change, note that as an option, the convertible backlite was offered in *folding glass*. A clear vinyl convertible backlite remained standard equipment.

Yet another styling change was made to

On '67—68 fastbacks, roofline continued to rear edge of decklid. Photo by Ron Sessions.

the side sculpture. Now, it took on the appearance of a two-scoop air intake. The entire sculpture of the side was flared, making it more predominant than earlier models. As a result, doors, fenders, hoods and trunk lids no longer interchanged with previous years, but the front clip and doors remained interchangeable between models of that year.

Interior—For 1967, there was a new instrument panel, rounding from the top down in a convex "roll." This provided more kneeroom. The instrument cluster changed to a five-dial arrangement, with two large dials placed under three smaller ones. The large dials, directly above the steering column, contained the 120-mph speedometer and a combination oil-pressure gage/ammeter. The three smaller units contained the fuel gage, optional clock, and temperature gage.

If you selected a tachometer, this was located in the spot formerly reserved for the oil-pressure gage/ammeter. These were replaced with warning lights on the face of the tachometer. Normally, the tach was calibrated to 6000 rpm. If you chose the HiPo 289, the tach was calibrated to 8000 rpm.

The steering wheel was changed to a deep-dish design. Safety pundits in Washington were beginning to pressure Detroit to provide more passive occupant crash

protection. Inside the cars, this manifested itself in soft trim for window cranks, door handles, a padded instrument panel, inertia-reel front seat belts, and rounded edges on all trim. In the center of the deep-dish steering wheel was a large padded horn button. The three-spoke wheel and horn button, or hub, were designed to collapse under impact.

In 1967, the pony interior, as such, was dropped. An Interior Decor group was provided as an option, but there were no more wild mustangs at full gallop across the back of the seat. A roof-mounted console with two map lights was the center of attention for this package. The luxury Interior Decor option included molded armrests in the door panel. This was set off with a brushed-aluminum trim panel. Courtesy lights were provided within the door. The instrument panel and center console were likewise decorated with brushed-aluminum trim, carrying the theme throughout the car.

With the hatchback, the package included padded rear-quarter trim panels. Of course, special upholstery trim was part of the package.

Desirable options included a roll-top center storage compartment and a tilt-away steering wheel to aid driver ingress and egress.

GT Equipment Group—The GT options were similar in 1967 to what they were in

1966, but with one exception: If the car was equipped with an automatic transmission, the side-stripe designation was GT/A (automatic) rather than GT.

Drive Train—For 1967, the 390-CID big-block engine was added to the existing lineup of three 289-CID V8s and the 200-CID six. This new 320-HP, 390-CID four-barrel came with either the four-speed manual transmission or the C-6 Cruise-O-Matic. A heavy-duty three-speed manual transmission was available at extra cost on the smaller engines.

The New Shelby—The big noise out of Detroit was the new Shelby G.T. 500. The G.T. 350 remained, but the 500 gathered the limelight. When the stock Mustang was widened to admit the larger 390, why not go all the way, thought Shelby, and squeeze in a 428. This added almost 50 more horsepower, bringing the rating to 355 HP. And the Paxton supercharger was again available on the G.T. 350. These, and the smaller G.T. 350s were the last of the Shelbys to be produced at the California plant. Everything was later moved to Ionia, Michigan, to the A. O. Smith assembly plant.

Shelby did some serious modifying in 1967. He began by lengthening the stock Mustang 3 in. in front with a fiberglass nose and longer hoodline. The new nose included a black-anodized expanded sheet-metal grille, housing two round 7-in.driving lights placed side by side in the grille center (although some states required that the lights be spaced farther outboard). Although the front bumper was stock '67 Mustang sans bumperettes, the lower front valance was a special fiberglass piece with a large air intake for the radiator.

At the rear, dechromed Cougar sequential taillamps were used. Similarly, a fiberglass decklid and quarter-panel extensions included a rear spoiler.

Three functional air scoops were added to the car. This included a hood scoop and a scoop replacing the side sculpture. This fed air to the rear wheel well rather than directly to the brakes. Little, if any, brake cooling was realized. Finally, the side quarter window was removed and a racy-looking air-extractor scoop was put in its place.

1968

Mustang had the lions share of the 1967 muscle-car market (Camaro, Cougar, Barracuda and Firebird) so little change was given to something that was working right. Ford would beef-up the engine-option list again, adding the 428-CID V8, and thereby giving Mustang two big-block

Side-marker lamps first appeared on '68 models. Photo by Ron Sessions.

Desirable options on '67—68 models included tilt-away column and roll-top console compartment, as shown on David Finner's stunning 1968 fastback. Photo by Ron Sessions.

engine options. Shelby would add a beautiful convertible to their line, increasing customer selection. Finally, some special-edition cars would be brought out. These are discussed later. For now, let's look more carefully at the changes Ford incorporated for 1968.

Body Styling—The grille bars were removed again, but the mustang was retained, albeit a smaller one. The sheet metal remained the same as that of the '67, and was interchangeable. The side-sculptured air scoops were removed and replaced with a thin, decorative chrome strip. Ford dropped the Exterior Decor Group in 1968 but left the louvered hood as an option. If you chose this option, you could also have a two-tone paint job: flat-black enamel from the louvers to the windshield.

As for the windshield, this was the first year of the federally mandated break-away rear-view mirrow, with the mirror base glued to the windshield.

The rear-end treatment remained basically the same as in '67. The quickest way to tell a '68 from a '67 is to look for the federally mandated side-marker lamps—'68s have them and '67s don't.

Interior—Very little was changed in the interior for '68. The instrument panel had the same instruments and configuration. The large gage on the right now incorporated a fuel gage where the oil-pressure gage was in 1967. The oil-pressure gage was transferred to the small, upper-left pod. If you elected for the 8000-rpm tachometer, it displaced both the fuel gage and ammeter. These became warning lights at the bottom of the tach as they were in '67.

The steering wheel was changed quite dramatically. The large, padded horn button was removed and replaced with a heavily padded, two-spoked horn bar—part of a continuing effort to increase interior safety, but improve appearance.

Seat-cover styling remained much the same. The Interior Decor Group, however, changed a little. Woodgrained veneer appliques were fixed to the instrument panel. The door panels became two-tone with a molded armrest, woodgraining in the center, a door pull, and lower grille with a red and white safety/courtesy lamp. Further, chrome trim on the brake and clutch pedal and fancy rectangular chrome buttons in the seat backs. With the coupe or fastback, you could have padded rear-quarter trim panels were part of the package. The overhead console remained for '68. Added though, was woodgraining in the steering-wheel horn bar and a vinyl grip on the T-handle shift lever.

GT Equipment Group—In 1968, the GT option included new GT "C" stripes, decorating the outside edge of the side sculpture. These replaced the 1967 stripes above the rocker panels. New also for '68 were GT chrome wheels. These were also finished between the chromed areas with reflective paint. If your car was equipped with a 390 or 427, some engine components such as the valve covers, oil-filter cap and air cleaner were chromed. Remaining from 1967 were the GT emblem, fuel cap, fog lights (minus the grille bar), low-restriction dual exhaust, F70x14 Wide Oval tires and heavy-duty suspension.

Sports Trim Group—This was a new option for 1968 and included: bright-metal trim around the wheel openings, two-tone hood (with louvered-hood option), special Argent (silver) painted wheels, Wide Oval raised-white-letter tires, woodgrained instrument cluster and knitted-vinyl inserts in the seats of the coupe and fastback. The list of Mustang options was growing by leaps and bounds.

Drive Train—Ford got real busy in 1968 with their engines. This was the first year of federally mandated emission controls in the 49 states. The standard 200-CID six-cylinder was retained, but Ford dropped the 289-CID four-barrel and 289 HiPo engines. The four-barrel 289 was replaced with a new 302-CID four-barrel small-block—basically a stroked 289. And by the middle of the production run, the 289-CID two-barrel was dropped and the void filled with a 302-CID two-barrel. High Performance engines were a must, so Ford offered two of them: a High Performance 302 and a 427-CID big-block. These both saw very limited production and were dropped in favor of the potent 428 Cobra Jet (CJ). In November of '67, Ford added the 250-CID six-cylinder.

Transmission choices included a stand-

Early '68 G.T.500s were available with 427-CID V8. Photo by Steve Christ.

One of the few '68 California Specials still on the road.

ard three-speed with the six-cylinder engines and 302 V8s. Four-speed transmissions were optional on the 302s, the remaining 289-CID two-barrels and the big-block 390. Cruise-O-Matic was an option on all engines, with the C-4 chosen for sixes and small-block V8s and the beefier C-6 used behind the big-blocks.

Shelbys—As stated earlier, Shelby manufacturing was moved to Ionia, Michigan. This was the result of two problems: first, Shelby was not able to find enough capacity to make quality fiberglass in the L.A. area and second, his lease was running out on the Shelby-American facility. Further, the name of the cars were changed to Shelby Cobra G.T.350 and G.T.500.

The biggest change was the addition of a convertible to the 350 and 500 series Shelbys. In mid-year, Mustang unveiled its new 428 Cobra Jet engine and Shelby introduced the G.T.500KR (King Of The Road). Exterior changes were limited to minor details in the grille area, hood scoop, and side vents. The '67 Cougar taillights were dropped in favor of 1965 Thunderbird taillights. Rocker-panel stripes carried the G.T.350, G.T.500 and G.T.500KR logos.

The big change in the Shelby interior was a new roll bar, particularly attractive in the new convertible line. Additional interior features included two Stewart-Warner gages in the center console: oil pressure and an ammeter.

Shelby engines for 1968 included the introduction of a 250-HP 302 four-barrel for the G.T.350. G.T.500s were equipped with a 360-HP 428 "Police Interceptor." Perhaps 50 G.T.500s were equipped with

the 400-HP, medium-riser 427. With a C-6 automatic transmission, these cars were special-order units. And we have it on good authority that Ford slipped a few 390 GT engines into G.T.500s due to a shortage of 428s.

The G.T.500KR replaced the 500 series mid-year and employed the same 428CJ as its sister Mustangs.

Special Edition Vehicles—In 1968, Ford introduced three Special Edition cars. These included the Mustang Sprint, California Special and the High Country Special.

The Mustang Sprint was offered in both a six- and eight-cylinder package. The six-cylinder model had GT stripes, full wheel covers and a pop-open gasoline cap. The eight-cylinder model had all of these items plus GT fog lights, styled steel wheels and Wide Oval tires.

The California Special was an attempt to further stimulate the already very active California market. These were marketed by the Southern California Ford Dealers and included such items as fog lights mounted to a chrome-free grille—no running mustang. It also included non-functional, fiberglass side scoops and a Shelby convertible fiberglass trunklid and rear-quarter extensions that formed a rear spoiler. Used also was a fiberglass rear panel with the same '65 Thunderbird taillights as found on the Shelby. Graphics included side stripes with a GT/CS logo on the side scoops and rear-deck stripes around the edge of the spoiler. The hood was retained with twist-type locks. California Specials bring a little more than a standard Mustang.

Note the distinctive Shelby-type simulated rear-brake air scoop. Shelby rear spoiler and taillamps were also used on the California Special.

The High Country Special was the Colorado Ford Dealers response to the California Special. Actually, the two "specials" were identical except for the logos. The GT/CS was replaced by a triangular shield incorporating a running mustang and the words High Country Special, 68. Both cars were available as coupes only.

1969

1969 was a year of radical styling changes. Probably more sheet metal changed between 1968 and 1969 than in all

By '69, the Detroit muscle-car horsepower wars had escalated with numerous large-displacement offerings. Here's the '69 Mach I with a 428 Cobra Jet. Photo courtesy of Ford Motor Company.

Standard fare in the '69 Mach I was the all-new 351-CID Windsor small-block V8. This SportsRoof sports SportsSlats (backlite louvers by any other name) as well as add-on front and rear spoilers.

of the preceding years. Ford was up to its corporate ears in the muscle-car race. The swoopy looking body designs did not belie the powerful machinery available under the hood. So competitive did Ford become with their muscle cars that the more-expensive Shelby line could no longer compete. Thus, 1969 was the last full production year for the Shelby.

Body Styling—Mimicking the '65—68 GT package, the front-end styling of the '69 Mustang featured quad headlamps, with the high-beam-only center lamps relocated to the outboard edges of the grille. These inner lights were mounted in a new black plastic grille with the Mustang logo offset to the left. The valance panel was changed to admit fresh air to the engine compartment.

The concave sculpted side of the car, a predominant feature since its inception, just about disappeared. In fact, this area of the body was made convex from front to back. The upper bounds of this styling feature was a body line from the lip of the headlight to the rear quarter panel. A three-section, reversed scoop at the end of the body line suggested an air exhaust vent on the coupe and convertible.

The fastback, now called the *Sports-Roof*, had the same body line. The body line, however, terminated at a non-functional air intake scoop at the front of the quarter panel immediately behind the door handle. Another body line extended from just above the rear bumper along the

rear quarter panel to just behind the rear wheel well to define the lower bounds of the convex body sculpture.

The cars grew in length as well as in performance. By adding 4 in. to the body, Ford increased the length of all three models to 187.4 in., the longest Mustang built up to that time.

The rear exterior somewhat resembled the concave style of 1968. The SportsRoof, however, had a spoiler built in to the deck lid and quarter-panel extensions.

Interior—A complete interior styling change accompanied the exterior changes. The instrument-panel dash pad was now a two-pod design, accentuating and dividing the driver and passenger compartments, giving a "cockpit" effect. Gages were deeply recessed into the instrument panel, with an overhanging crash pad greatly reducing reflected glare from the instrument faces.

The new instrument cluster featured four gage pods: two large ones centered behind the steering wheel and two smaller ones at either side. From left to right, these were the ammeter, speedometer with high-beam indicator, fuel and temperature gage pod with two warning lights for seat belts and emergency brake, and the oil-pressure gage.

Two interior groups were available: Interior Decor Group and Deluxe Interior Decor Group. The Deluxe Interior Decor Group was available only on the convertible and SportsRoof. At additional cost,

you could have high-back bucket seats. The biggest difference between the standard and deluxe interiors was the wood-grained, vinyl applique on the dash. To this was added a clock mounted above the glove compartment. A tachometer, if so ordered, replaced the oil-pressure gage and coolant-temperature gage. These became warning lights. Adjustable front-seat headrests were offered on low-back buckets for the first time in 1969.

GT Equipment Group—1969 was the last year of the GT equipment group and, as such, offerings were basically a carryover from 1968. Ford was now putting more emphasis on selling *packaged* cars such as the Mach I and the Grande—cars with elements of the GT option group. The few cars sold with the GT option are among the rarest of Mustangs.

Drive Trains—For 1969, Ford dropped the 302 four-barrel engine, but added two- and four-barrel versions of a new small-block V8, the 351—basically a stroked 302. This provided the customer with eight engine choices: a standard 200-CID six and optional 250-CID six; three small-block V8s, the 302 two-barrel, 351 two-barrel and four-barrel; and two big-block V8s, the 428 and 428 CJ (Ram-Air).

Three-speed manual transmissions were offered as standard on the two sixes and the three small-blocks. The four-speed boxes were available as an option on all engines except the two sixes. A Cruise-O-Matic was again an option on all eight engines.

Shelby-ized 428CJ with Ram-Air found its way into '68, '69 and retitled '70 Shelby G.T.500s. Photo courtesy of Ford Motor Company.

Not all of the supercar action was with big-block V8s. Here's the '69 Boss 302 Mustang. Photo by Ron Sessions.

For '69, Boss 302 engine was rated at 290 HP in street trim. Photo by Ron Sessions.

Specialty Models—Specialty models were the really big news for 1969. This year introduced the Mach I, Boss 302 and 429, and the luxurious Grande.

Mach I—The Mach I was developed on the new fastback SportsRoof and made

available with any of five V8 engines. It featured a deluxe interior with high-back bucket seats, a deluxe three-spoke steering wheel, console with woodgrain vinyl trim and special carpet.

Outside was added a non-functional hood scoop, low-gloss black—matte finish—painted hood, racing-style hood pins, color-keyed side mirrors, and Mach I body stripes in one of three color combinations: black with gold, red with gold or gold and white. Chrome styled-steel wheels, pop-open gas cap, dual exhaust with four chrome outlets (four-barrel en-

gines only), E70x14 tires and rocker-panel moldings completed the exterior treatment of the performance-oriented Mach I.

Boss 302—The Boss 302 was designed to compete with Chevrolet's Camaro Z/28 in SCCA's Trans-Am race series. Available in the fastback Sports-Roof only, its powerplant gave it its name, the 290-HP 302-CID V8 with 351C style cylinder heads. Similar to the '65 and '66 HiPo models, it had a four-speed top-loader transmission, 9-in. rear axle, stiffer suspension and 16:1 steering, to name a few.

The Boss 302 was basically a factory

Boss 429 engine was a tight squeeze in the '69—70 Mustang. Shock-tower clearancing and other mods were necessary to shoehorn the big mill in. Photo by Roger Huntington.

Ford's biggest gun ever, the 375-HP Boss 429, was dropped into a select number of '69 and '70 Mustangs to legalize the engine for NASCAR use in Grand National Torinos. Photo courtesy of Ford Motor Company.

road racer with muscle-car styling. This styling included elimination of the Sports-Roof's simulated side scoop, a C-shaped stripe with "Boss 302" lettering, functional front spoiler and black-out hood, rear deck and back panel. You could have an optional set of black, rear-window louvers and adjustable, stand-up rear airfoil over the tail. Four vibrant colors were available: bright yellow, calypso coral, wimbledon white and acapulco blue.

The 302-CID HO (high-output) engine was available in both racing and street versions. The street version was rated at 290 HP at 5800 rpm and 290 ft-lb of torque at 4300 rpm. The beefier racing engine had close to 450 usable HP.

Canted-valve cylinder heads breathed through a high-rise intake manifold and 780-cfm Holley four-barrel for the street version. The heads had extra-large valves (2.33-in. intake and 1.71-in. exhaust) with solid valve lifters and a high-lift camshaft.

Internal beefing for this wild street engine was similar to the HO 429. This included oversized forged four-bolt main-bearing caps, forged connecting rods, caps and bolts and forged, extruded aluminum pistons. Every 302 forged-steel crankshaft was electronically balanced, both statically and dynamically. The exterior was completed with chrome or cast-aluminum rocker covers and a high-capacity, dual-point ignition system. This was all matched to a four-speed manual transmission (with ratios specially spaced

to match the 302's output curve) and a 3.50:1 rear axle. An option was a locker rear axle with either 3.50, 3.91 or 4.30 ratios.

The racing version had special machining for *dry-decking* rather than gaskets, special intake manifold with individual runners, two four-barrel Holleys rated at 1100 cfm *each*, and fabricated tubular exhaust headers tuned for max performance. A cast-aluminum oil pan was fitted with a windage tray to prevent oil splash. To improve high-rpm lubrication, an Indy-style, cross-drilled, forged-steel crankshaft was added. Valves were forged steel with chrome-plated hollow stems and tulip-shaded heads. The exhaust valves were sodium-filled for maximum cooling.

Additionally, the racing Boss 302 included wide-shouldered connecting rods with beefed-up bolts. These were similar to those developed for the 289 LeMans winning GT-40s.

Chassis refinements on the street car included: weight distribution of 56/44 front-to rear, heavy-duty suspension with staggered rear shocks, 16:1 quick-ratio steering with 3.74 turns lock-to-lock, F60x15 fiberglass-belted tires on 7-in. Magnum 500 steel wheels and floating-caliper power front disc brakes with ventilated cast-iron discs.

Boss 429—Continuing Ford's push for racing, the Boss 429 met NASCAR's homologation rule for Grand National racing. This rule stated that at least 500 units

of a particular engine *or* body style must be manufactured and sold to the public. To squeeze the big Boss 429 (*Blue Crescent*) into the engine compartment, special front-suspension spring towers had to be installed, making the engine compartment wider. This and other engineering changes were accomplished by Kar Kraft, Ford's in-house race-car fabrication and preparation arm, in their small Brighton, Michigan assembly facility. The exterior featured a functional hood scoop, a front spoiler, a Boss 429 decal on the front fenders and dual color-keyed mirrors. Performance features included the Boss 429 engine, four-speed transmission, 3.91 Traction-Loc differential, power disc brakes in front, heavy-duty suspension, power steering, 65-amp alternator, and an 85-amp battery mounted in the trunk.

Grande—Not all of Ford's effort went into performance. The Grande bears witness to this fact. This was Ford's answer to the buyer who wanted T-Bird luxury in a smaller, sportier package. Available as a coupe only, it provided as standard equipment the Deluxe Decor Interior, plus simulated woodgrain instrument-panel trim, deluxe three-spoke steering wheel, special door panels, padded interior quarter panels, and special cloth-inserted bucket seats.

Outside, wire wheel covers, dual mirrors, C-pillar logo, duo-toned paint stripe under the fender line, and a chrome trim molding around the fender well set the Grande apart from other Mustang coupes.

There are five, count 'em, five air scoops on this Shelby hood. Originally, air scoops were functional and critical for cooling and breathing. By 1969, however, five scoops on the hood were styling decisions rather than necessities. Photo by Tom Monroe.

On '69—70 Shelby, note center fuel fill behind license plate and controversial center-exit exhaust.

Shelby—The year 1969 ended up being the last full production year for the Shelby. And, as the Mustang was restyled, so was the Shelby. NACA air scoops were in vogue for '69. Each car was equipped with no less than nine! Five scoops decorated the hood while each wheel well had its own cooling scoop—one in each front fender and rear-quarter panel. Big news was the all-fiberglass front end—fenders, hood and front valance.

The all-fiberglass front-end was a "trial balloon" for the 1971 Mustang's front-end styling. Full-width taillamps were retained at the rear and a unique exhaust outlet was used. It exited at the center immediately below the fuel-tank filler. This made for some excitement soon after the fuel tank was filled. Fuel that spilled over or vapors vented from the tank ran down into the exhaust outlet, sometimes catching on fire. Going down the road, the car looked like a rocket, what with two long sheets of flames trailing the car.

Shelby once again offered a convertible and the G.T.350 and G.T.500 series. The KR designation was dropped because the 428 Cobra Jet became standard in the G.T.500. The 351 Windsor engine powered the G.T.350.

1970

In 1970, Ford met the goals it had set. Mustang was the number-one selling *pony car*, well ahead of Chevy's Camaro. The

A fully restored, prize-winning example of the '69 Shelby.

Boss 302, under the skillful technique of Parnelli Jones and George Follmer, swept the Trans-Am series, and all was well in Fordland. Well, almost all. Unfortunately, the Shelby did not make production in 1970. The few Shelbys sold were dealer-prepared leftover '69s. The Boss 302 and 429 were simply too formidable competitors, and overall, the performance market was not receiving much attention due to government safety and emission-control regulations.

Body Styles—A few cosmetic changes were made for 1970. Simulated air scoops were added to the die-cast end caps at the front of the fenders. Dual headlights were dropped in favor of single headlights mounted in a wider grille. The running pony raced from its '69 location on the driver's side of the grille, back to the center. Other changes included optional Sports Slats for the SportsRoof model, a plastic spoiler, dual accent paint stripes and a new vinyl roof.

A '70 vintage Boss 302. On all '70 Mustangs, headlights were moved into outboard ends of grille. End caps became simulated air scoops. Photo courtesy of Ford Motor Company.

Interior—The interior, too, stayed basically the same as the '69. However, *highback* bucket seats which were optional in '69 became standard in '70. A new shape to the steering wheel accented a new steering column with the ignition switch mounted thereon for the first time. There was only one decor option for 1970. It was the same as the '69 Deluxe Interior Decor.

Drive Trains—The biggest change in the drive train came about by dropping the 351W for the beefier 351C. Although some modification was made in the blocks, a comparison of the 351 Windsor and 351 Cleveland is basically a comparison of the heads. The 351C four-barrel engine generated 300 HP at 5400 rpm. Its Windsor two-barrel counterpart was rated at 250 HP at a lower 4600 rpm.

To achieve these increases, new canted-valves with larger heads were designed for the 351C. Additionally, a four-barrel carburetor was mounted, feeding through larger ports. Spent exhaust was also removed through larger ports. Valve size for the Windsor series was 1.84-in. intake and 1.54-in. exhaust. These were increased on the Cleveland to 2.18-in. intake and 1.70-in. exhaust.

Another performance option was the Drag Pac. Available only with the 428CJ was a Traction-Lok differential with a 3.91:1 final drive, or a Detroit Locker with a higher 4.30:1 ratio. An engine oil cooler was also part of the package.

Mach I—The Mach I was toned down a bit for 1970. Gone were the side stripes and fully blacked-out hood. A flat-black graphic replaced the all-black hood. Within the two stripes running along the side of the non-functional hood scoop was the engine displacement (302, 351, or 428). The Styled Steel wheels were replaced with mag-style wheel covers. A stripe and honeycomb rear panel rounded out the major changes.

Boss 302—The Boss 302 grew more radical in appearance. An optional "shaker" hood and side graphics were added, as were color-keyed mirrors and flat-black rear panel. For 1970, 15-in. wheels with new flat hub caps and beauty rings were standard. Magnum 500 Wheels were available as an option.

Boss 429—The Boss 429 for 1970 remained much the same as it was in 1969, the major change being a black-painted hood scoop. The sheet-metal changes of the '70 SportsRoof were reflected in the Boss 429.

Grande—For 1970, the Grande remained the luxury flagship of the Mustang line. Added this year was a new landau vinyl roof and a special cloth insert in the seat upholstery.

Grabber SportsRoof—A few Sports-Roofs, dubbed *Grabbers*, were upgraded with a fancy reflective C-stripe graphic with the legs of the "C" pointing to the rear of the car from behind the front wheel well.

The Grabber was also equipped with hubcaps and beauty rims. Another version sported a graphic similar to the one on the Boss 302. At the top of the graphic, just below the side mirror was a 302 or 351 designation.

PONYCAR EPITAPH

1970 can be considered the last year of the muscle car. Federal regulations for safety and emissions buried Detroit under reams of bureaucratic forms, red tape and re-engineering.

Devastating insurance costs to the owners of these powerful machines made the potential buyer think twice (as is the case today) before investing in a car that could add thousands of dollars to the annual cost of car ownership. 1970 was also the last year Ford first competed in the Trans-Am Series; turning its attention away from speed and power to what it felt the public wanted: big, heavy luxury cars.

The result were the styling misadventures of 1971—73. The Mustang grew fat and lazy. The '71 was extended 2 in. and blimped up by 600 lb. Styling too, took a nosedive. Consider what *Car Life* said in September, 1970 of the '71 Sports-Roof. "The fastback or SportsRoof is actually a flatback. The roof angle is only fourteen degrees. The rear window would make a good skylight. A glance in the rearview mirror provides an excellent view of the interior with a small band of road visible near the top of the mirror."

By 1972, even the pretense of performance cars was gone. The 429 powerplants were dropped and all of the remaining engines were detuned to run on low-octane regular fuel. Nevertheless, 1973 was a good year. Mustang sales rebounded that year, but it was clear that the 1971—73 models had strayed far from the original concept. Waiting in the wings, however, was a small, more fuel-efficient 1974 Mustang—the Mustang II.

Lee Iacocca thought the Pinto based Mustang II would recapture the spirit of the 1965 Mustang. It is an old axiom that the camel is a horse designed by a committee. In some respects, that is the story of the Mustang II.

The design firm of Ghia, in Turin, Italy produced the first prototype design. Next, a competition was held between Ford's own designers. These various designs were then shown to select groups across the U.S. It was a draw between the fastback and notchback, but Ford did not want to produce two body styles. By production time, a fastback model was ready to be built. However, at the last minute, Iacocca decided that a

Ford warranty, or identification, plate at rear of driver's door gives valuable information on the car's ancestry and equipment. Photo by Ron Sessions.

By all means, tag on door should match this number stamped into both front inner fenders adjacent to hood opening. This stamping is visible with fender attached on left side; right fender must be removed to see stamping on that side. Photo by Ron Sessions.

On Shelby Mustangs, ID plate is riveted to top of left inner fender. Photo by Ron Sessions.

notchback must be included in the lineup for '74. Designers were then faced with the problem of making a notchback roof fit a fastback body and making common doors work with both roof lines.

The car that resulted was a pieced-together assembly of parts on a Pinto chassis. Advertising promotions said the Mustang II would sell at less than a $3000 base price.

What resulted at introduction was a resounding thud. There were no base models available at $2895. The only thing the buyer found were option-laden models at $4000 to $4500. Ford had predicted sales of over 31,000 the first month; 18,000 materialized.

What saved the Mustang II were the Arabs. When the oil was shut off in October, 1973 gas-hungry Americans flocked to the compact-car market carrying the Mustang II to record sales figures. By 1976, the car's popularity had peaked, this nation's fuel panic had subsided somewhat and Ford was working on a third-generation Mustang that would be introduced in 1979. It's history that the '79 Mustang and mid-'80s GT and SVO permutations put the pony/muscle car back on track.

VEHICLE IDENTIFICATION

For more information about your particular car, check the vehicle identification number (VIN). This eleven-digit, alphanumeric number is stamped on the car's warranty plate or certification label on the rear face of the driver's door. From 1965 to 1969, it was a reverse-stamped aluminum plate. From 1970 on, it became a paper tag with a vinyl overlay.

The VIN was also stamped into the top upper flange of the left front fender apron and, starting in 1968, stamped on an aluminum plate riveted to the top of the instrument panel, visible through the windshield. In 1968, the plate was located on the passenger's side at the junction of the panel and windshield. In 1969 and thereafter, it was moved to the same location on the driver's side.

Model Year
The model year number indicates the second digit of the year: 6(5), 6(6), 6(7).

Consecutive Unit Number
Each plant, each year beings numbering the vehicles with the number 100001. Other models such as Falcons and Fairlanes were mixed into the assembly line along with the Mustangs. Therefore, if your unit number reads 100004, it means your car was the fourth built, not the fourth Mustang built.

T-5

Had you purchased a '65 or '66 Mustang in West Germany, there would be nothing on the vehicle to show it was a Mustang. In fact, you would not have purchased a Mustang, but rather a *T-5*. One German manufacturer had already claimed the name "Mustang" for its own. Rather than pay a very stiff royalty fee, Ford chose to delete all reference to that name, calling it instead by its original Ford Motor Company project designation, the T-5.

So, the name Mustang disappeared from the front-fender script, horn ring, wheel covers and gas cap. Other changes included the use of heavy-duty suspension with a 1-in. front stabilizer bar, engine-compartment *export brace,* and a 200 kilometer speedometer.

The Mustang that wasn't, the T-5. This catchy moniker was Ford's internal engineering designation for the Mustang and appeared on all Mustangs sold in West Germany in '65—66. Photo by Ron Sessions.

T-5s were fitted with, among other things, this special metric speedometer, calibrated up to 200kmh—that's kilometers per hour, not miles per hour, and this is a six-banger! Photo by Ron Sessions.

By studying the nearby example and accompanying chart, you can see what each letter or number indicates, such as body style, interior and exterior trim, and other pertinent information.

1965—67 Shelbys had a serial-number tag riveted over the regular Ford VIN number on the driver's side inner-front-fender panel. In 1968, the Shelby and Ford numbers were combined and located in the same place. the middle of 1968, the Shelby had a regular Ford warranty plate but with the added stamping, SPECIAL PERFORMANCE VEHICLE. Because the '70 Shelbys were really only leftover '69 models updated by dealers with a new chin spoiler and black hood stripes, the only change was in the first digit, changing the 9 to a 0.

BODY STYLE

	65	66	67	68	69	70
Fastback	63	63	63	63	N/A	N/A
Hardtop	65	65	65	65	65	65
Convertible	76	76	76	76	76	76
Sportsroof	N/A	N/A	N/A	N/A	63	63
Standard Int.	A	A	A	A	A	A
Luxury Int.	B	B	B	B	B	B
Special Int.	N/A	N/A	N/A	N/A	E+	E+
Bench seats	C	C	C	C/D*	C/D*	C/D*

*Deluxe
+Grande

BODY SERIAL CODE

	65	66	67	68	69	70
Fastback	09	09	02	02	-	05 (Mach I)
Hardtop	07	07	01	01	01	01
Convertible	08	08	03	03	03	03
Sportsroof	-	-	-	-	02	02

EXTERIOR PAINT

CODE	64-1/2	65	66	67	68	69	70
A	Raven Black	Raven Black	Raven Black	Raven Black	Raven Black	Raven Black	Raven Black
B	Pagoda Green	Midnight Turquoise	—	Frost Turquoise	Royal Maroon	Royal Maroon	—
C	—	Honey Gold	—	—	—	Black Jade	Dary Ivy Green Met.
D	Dynasty Green	Dynasty Green	—	Acapulco Blue	Acapulco Blue	Acapulco Blue	Yellow
E	—	—	—	—	—	Aztec Aqua	—
F	Guardsman Blue	—	Light Blue	Arcadian Blue	Gulfstream Aqua	Gulfstream Aqua	—
G	—	—	—	—	—	—	Med. Lime Metallic
H	Caspian Blue	Caspian Blue	Sahara Beige	Diamond Green	—	—	—
I	Champagne Beige	Champagne Beige	—	Lime Gold	Lime Gold	Lime Gold	—
J	Rangoon Red	Rangood Red	—	—	—	—	Grabber Blue
K	Silversmoke Gray	Silversmoke Gray	Nightmist Blue	Nightmist Blue	—	—	Bright Gold Metallic
L	—	—	—	—	—	—	—
M	Wimbledon White	Wimbledon White	Wimbledon White	Wimbledon White	Wimbledon White	Wimbledon White	Wimbledon White
N	—	—	—	Diamond Blue	Diamond Blue	—	Pastel Blue
O	Tropical Turquoise	Tropical Turquoise	—	—	Seafoam Green	—	—
P	Praire Bronze	Praire Bronze	Antique Bronze	—	—	Winter Blue	—
Q	—	—	Brittany Blue	Brittany Blue	Brittany Blue	—	Medium Blue Met.
R	Ivy Green	Ivy Green	Dark Green Metallic	—	Highland Green	—	—
S	Cascade Green	—	—	Dusk Rose	—	Champagne Gold	Med. Gold Met.
T	—	—	Candyapple Red	Candyapple Red	Candyapple Red	Candyapple Red	Red
U	—	—	Tahoe Turquoise	—	Tahoe Turquoise	—	Grabber Orange

EXTERIOR PAINT

CODE	64-1/2	65	66	67	68	69	70
V	Sunlight Yellow	Sunlight Yellow	Emberglo	Burnt Umber	—	—	—
W	—	—	—	Clearwater Aqua	Meadowlark Yellow	Meadowlark Yellow	—
X	Vintage Burgundy	Vintage Burgundy	Vintage Burgundy	Vintage Burgundy	Presidential Blue	—	—
Y	Skylight Blue	Silver Blue	Silver Blue	Dark Moss Green	Sunlit Gold	Indian Fire	—
Z	Chantilly Beige	—	Med. Sage Metallic	Sauterne Gold	—	—	Grabber Green
1	—	—	—	—	—	—	Calypso Coral
2	—	—	—	—	—	Lime	Light Ivy Yellow
3	Polly Red	Poppy Red	—	—	—	Calypso Coral	—
4	—	—	Silver Frost	Silver Frost	—	Silver Jade	—
5	Twilight Turquoise	—	Signalflare Red	—	—	—	—
6	—	—	—	Pebble Beige	Pebble Beige	Pastel Gray	Silver Blue Met.
7	Phoenician Yellow	—	—	—	—	—	—
8	—	Springtime Yellow	Springtime Yellow	Springtime Yellow	—	—	—
Special	—	—	Med. Palomino Metallic	Playboy Pink	—	—	—
	—	—	Med. Silver Metallic	Anniversary Gold	—	—	—
	—	—	Ivy Green Metallic	Columbine Blue	—	—	—
	—	—	Tahoe Turq. Metallic	Aspen Gold	—	—	—
	—	—	Maroon Metallic	Blue Bonnet	—	—	—
	—	—	Silverblue Metallic	Timberline Green	—	—	—
	—	—	Sauterne Gold Met.	Lavender	—	—	—
	—	—	Light Beige	Bright Red	—	—	—

INTERIOR TRIM CODES

1964-1/2
42	White Vinyl w/ Blue Trim
45	White Vinyl w/ Red Trim
46	White Vinyl w/ Black Trim
48	White Vinyl w/ Ivy Gold Trim
49	White Vinyl w/ Palomino Trim
56	Black Vinyl & Cloth
82	Blue Vinyl w/ Blue Trim
85	Red Vinyl w/ Red Trim
86	Black Vinyl w/ Black Trim
89	Palomino Vinyl w/ Palomino Trim

1965
22	Blue Vinyl w/ Blue Trim
25	Red Vinyl /w Red Trim
26	Black Vinyl w/ Black Trim
28	Ivy Gold Vinyl w/ Gold Trim
29	Palomino Vinyl w/ Palomino Trim
32	Blue Bench
35	Red Bench
36	Black Bench
39	Palomino Bench
62	Blue & White (Luxury)
65	Red & White (Luxury)
66	Black & White (Luxury)
67	Aqua & White (Luxury)
68	Ivy Gold & White (Luxury)
69	Palomino (Luxury)
76	Black Fabric & Vinyl
79	Palomino Fabric & Vinyl
D2	White /w Blue Trim
D5	White /w Red Trim
D6	White w/ Black Trim
D8	White w/ Ivy Gold Trim
D9	White w/ Palomino Trim
F2	White /w Blue Trim (Luxury)
F5	White w/ Red Trim (Luxury)
F6	White w/ Black Trim (Luxury)
F7	White w/ Aqua Trim (Luxury)
F8	White w/ Ivy Gold Trim (Luxury)
F9	White w/ Palomino Trim (Luxury)

1966
22	Blue w/ Blue Trim
25	Red w/ Red Trim
26	Black w/ Black Trim
27	Aqua w/ Aqua Trim
32	Blue Bench
35	Red Bench
36	Black Bench
62	Blue & White (Luxury)
64	Emberglo & Parchment (Luxury)
65	Red (Luxury)
66	Black (Luxury)
67	Aqua & White (Luxury)
68	Ivy Gold & White (Luxury)
C2	Parchment w/ Blue Trim Bench
C3	Parchment w/ Burgundy Trim Bench
C4	Parchment w/ Emberglo Trim Bench
C6	Parchment w/ Black Trim Bench
C7	Parchment w/ Aqua Trim Bench
C8	Parchment w/ Ivy Gold Trim Bench
C9	Parchment w/ Palomino Trim Bench

D2	Parchment w/ Blue Trim
D3	Parchment w/ Burgundy Trim
D4	Parchment w/ Emberglo Trim
D6	Parchment w/ Black Trim
D7	Parchment w/ Aqua Trim
D8	Parchment w/ Ivy Gold Trim
D9	Parchment w/ Palomino Trim
F2	Parchment w/ Blue Trim (Luxury)
F3	Parchment w/ Burgundy Trim (Luxury)
F4	Parchment w/ Emberglo Trim (Luxury)
F6	Parchment w/ Black Trim (Luxury)
F7	Parchment w/ Aqua Trim (Luxury)
F8	Parchment w/ Ivy Gold Trim (Luxury)
F9	Parchment w/ Palomino Trim (Luxury)

1967
2A	Black Standard/Buckets
2B	Blue Standard/Buckets
2D	Red Standard/Buckets
2F	Saddle Standard/Buckets
2G	Ivy Gold Standard/Buckets
2K	Aqua Standard/Buckets
2V	Parchment Standard/Buckets
4A	Black Bench
4V	Parchment Bench
5A	Black Comfortweave (Luxury)
5V	Parchment Comfortweave (Luxury)
6A	Black Buckets (Luxury)
6B	Blue Buckets (Luxury)
6D	Red Buckets (Luxury)
6F	Saddle Buckets (Luxury)
6G	Ivy Gold Buckets (Luxury)
6K	Aqua Buckets (Luxury)
6O	Parchment Buckets (Luxury)
7A	Black Comfortweave Buckets
7V	Parchment Comfortweave Buckets

1968
2A-6A*	Black Vinyl Buckets
2B-6B*	Blue Vinyl Buckets
2D-6D*	Red Vinyl Buckets
2F-6F*	Saddle Vinyl Buckets
2G-6G*	Ivy Gold Vinyl Buckets
2K-6K*	Aqua Vinyl Buckets
2V-6V*	Parchment Vinyl Buckets
2Y-6Y*	Nugget Gold Vinyl Buckets
7A-5A*	Black Comfortweave Buckets
7B-5B*	Blue Comfortweave Buckets
7D-5D*	Red Comfortweave Buckets
8A-9A*	Black Comfortweave Bench
8B-9B*	Blue Comfortweave Bench
8D-9D*	Red Comfortweave Bench
8V-9V*	Parchment Comfortweave Bench
50	Parchment Comfortweave Buckets
* Used with decor group	

1969
1A	Black Cloth & Vinyl (Luxury)
1B	Blue Cloth & Vinyl (Luxury)
1G	Ivy Gold Cloth & Vinyl (Luxury)
1Y	Nugget Gold Cloth & Vinyl (Luxury)
2A	Black Vinyl Buckets

2B	Blue Vinyl Buckets
2D	Red Vinyl Buckets
2G	Ivy Gold Vinyl Buckets
2Y	Nugget Gold Vinyl Buckets
3A	Black Knitted Vinyl, Mach I
3D	Red Knitted Vinyl, Mach I
3W	White Knitted Vinyl, Mach I
4A-DA*	Black Comfortweave Hi Buckets
4D-DD*	Red Comfortweave Hi Buckets
DW	White Comfortweave Hi Buckets
5A	Black Comfortweave (Luxury)
5B	Blue Comfortweave (Luxury)
5D	Red Comfortweave (Luxury)
5G	Ivy Gold Comfortweave (Luxury)
5W	White Comfortweave (Luxury)
5Y	Nugget Gold Comfortweave (Luxury)
7A	Black Convertible Deluxe Buckets
7B	Blue Convertible Deluxe Buckets
7D	Red Convertible Deluxe Buckets
7G	Ivy Gold Convertible Deluxe Buckets
7W	White Convertible Deluxe Buckets
7Y	Nugget Gold Convertible Deluxe Buckets
8A-9A*	Black Comfortweave Bench
8B-9B*	Blue Comfortweave Bench
8D-9D*	Red Comfortweave Bench
8Y-9Y*	Nugget Gold Comfortweave Bench

*Interior Decor Group

1970
3A	Black Knitted Vinyl Mach I
3B	Blue Knitted Vinyl Mach I
3E	Vermillion Knitted Vinyl Mach I
3F	Ginger Knitted Vinyl Mach I
3G	Ivy Knitted Vinyl Mach I
3W	White Knitted Vinyl Mach I
AA	Black Houndstooth Cloth & Vinyl
AB	Blue Houndstooth Cloth & Vinyl
AE	Vermillion Houndstooth Cloth & Vinyl
AF	Ginger Houndstooth Cloth & Vinyl
AG	Ivy Houndstooth Cloth & Vinyl
BA	Black Vinyl
BB	Blue Vinyl
BE	Vermillion Vinyl
BF	Ginger Vinyl
BG	Ivy Vinyl
BW	White Vinyl
CE	Vermillion Blazer Stripe Cloth
CF	Ginger Blazer Stripe Cloth
EA	Black Comfortweave Vinyl
EB	Blue Comfortweave Vinyl
EG	Ivy Comfortweave Vinyl
EW	White Comfortweave Vinyl
TA	Black Comfortweave Vinyl
TB	Blue Comfortweave Vinyl
TG	Ivy Comfortweave Vinyl
TW	White Comfortweave Vinyl
VE	Vermillion Blazer Stripe Cloth
VF	Ginger Blazer Stripe Cloth

DATE CODES

The first two digits of the code represent the day of the month: 01A = 1st of January. If the model extends beyond 12 months, a second-year code will be used.

Month	1st Year Code	2nd Year Code
January	A	N
February	B	P
March	C	Q
April	D	R
May	E	S
June	F	T
July	G	U
August	H	V
September	J	W
October	K	X
November	L	Y
December	M	Z

DSO (DOMESTIC SPECIAL ORDER) DISTRICT CODES

Appearing in this space is the two-digit number of the district which ordered the unit. Units built on special order, domestic or foreign, have the complete order number in this space.

11	Boston	45	Davenport
12	Buffalo	51	Denver
13	New York	52	Des Moines
14	Pittsburgh	53	Kansas City
15	Newark	54	Omaha
21	Atlanta	55	St. Louis
22	Charlotte	61	Dallas
23	Philadelphia	62	Houston
24	Jacksonville	63	Memphis
25	Richmond	64	New Orleans
26	Washington	65	Oklahoma City
31	Cincinnati	71	Los Angeles
32	Cleveland	72	San Jose
33	Detroit	73	Salt Lake City
34	Indianapolis	74	Seattle
35	Lansing	81	Ford of Canada
36	Louisville	83	Government
41	Chicago	84	Home Office Reserve
42	Fargo	85	American Red Cross
43	Rockford	89	Transportation Service
44	Twin Cities	90-99	Export

AXLE RATIO CODES

	65	66	67	68	69	70
1	3.00:1	3.00:1	3.00:1	2.75:1	2.50:1	—
2	—	2.83	2.83	2.79	2.75	2.75:1
3	3.20	3.20	3.20	—	2.79	2.79
4	3.25	3.25	3.25	2.83	2.80	2.80
5	3.50	3.50	3.50	3.00	2.83	2.83
6	2.80	2.80	2.80	3.20	3.00	3.00
7	3.80	—	—	3.25	3.10	3.10
8	3.89	3.89	—	3.50	3.20	3.20
9	4.11	—	—	3.10	3.25	3.25
A	3.00*	3.00*	3.00*	—	3.50	3.50
B	—	—	—	—	3.07	3.07
C	3.20*	3.20*	3.20*	—	3.08	3.08
D	3.25*	3.25*	3.25*	—	3.91	—
E	3.50*	3.50*	3.50*	3.00*	4.30	—
F	2.80*	2.80*	—	3.20	—	2.33
G	3.80*	—	—	3.25*	—	—
H	3.89*	3.89*	—	3.50*	—	—
I	4.11*	—	—	—	—	—
J	—	—	—	—	2.50*	—
K	—	—	—	—	2.75*	2.75*
L	—	2.83*	—	—	2.79*	—
M	—	—	—	—	2.80*	2.80*
N	—	—	—	—	2.83*	—
O	—	—	—	—	3.00*	3.00*
P	—	—	—	—	3.10*	—
Q	—	—	—	—	3.20*	—
R	—	—	—	—	3.25*	3.25*
S	—	—	—	—	3.50*	3.50*
T	—	—	—	—	3.07*	—
U	—	—	—	—	3.08*	—
V	—	—	—	—	3.91*	3.91*
W	—	—	—	—	4.30*	4.30*
X	—	—	—	—	—	2.33*
Y	—	—	—	—	—	—

*Equa-Lock Differential

TRANSMISSION CODES

	65	66	67	68	69	70
1	3-Spd Man	3-Spd Man	3-Spd Man	3-Spd Man	3-Spd Man	3-Spd Man
3	—	—	3-Spd Man	—	—	—
5	4-Spd Man	4-Spd Man	4-Spd Man	4-Spd Man	4-Spd Man	4-Spd Man
6	C-4 Auto	C-4 Auto	—	—	4-Spd Man*	4-Spd Man*
E	—	—	—	—	—	—
V	—	—	C-6 Auto	C-6 Auto	C-6 Auto	C-6 Auto
W	—	—	C-4 Auto	C-4 Auto	C-4 Auto	C-4 Auto
Y	—	—	—	—	FMX Auto	FMX Auto

*Close-ratio

ASSEMBLY PLANTS

F—Dearborn, MI
R—San Jose, CA
Y—Metuchen, NJ

ENGINE CODES

Model	65	66	67	68	69	70
170	V					
200	T	T	T	T	T	T
250					L	L
260 (V8)	F					
289 2V	C	C	C	C		
289 4V ('64-1/2 only)	D					
289 4V	A	A	A			
289 HiPerf	K	K	K			
302 2V					F	F
302 4V				J		
302 4V Boss					G	G
351 2V Cleveland or Windsor					H	H
351 4V Cleveland only					M	M
390 2V				Y		
390 4V			S	S	S	
427 4V				W		
428 4V				P		
428 4V CJ					O	O
428 4V SCJ					R	R
429 4V Boss					Z	Z

2

Buying A Used Mustang

You may not have to be quite as picky as this Southern Arizona Mustang Club judge Dave Carroll is at checking out this '65 notchback, but it pays to take your time. After all, it's your money. Photo by Ron Sessions.

One way to find a really excellent restoration candidate is to join a Mustang club or attend a Mustang club car show, such as this one, the Great Pumpkin Car Show held by the Southern Arizona Mustang Club every Halloween. Photo by Ron Sessions.

You are about to embark on one of the most exciting, fun-filled activities a person can become involved with: buying and restoring a car. And in this case, it's a Mustang. Fun-filled, yes, if you know what you're doing. But it could become a black hole for your time and hard-earned money if you don't know how to go about it. Hopefully, reading this section will ensure that you don't get in over your head and your project will come out just the way you planned.

Planning is the key word here. You must have a plan, well developed before you begin, and then you must stay with that plan. The plan will include your budget, your time and your abilities. If you shift any one of these, you court disaster. Each is tied to the other. If you run out of time, it will probably cost you more money. If you overextend your abilities, you'll either need more time or more money to get back on track. So, determine your budget, time and abilities before you begin. Then figure out just what kind of project car you want.

WHAT TO BUY?

There are two types of cars in the hobby: *show cars* and *drivers*. Often there is an overlapping of these two types. This cross between the two is sometimes called a "parade car." And yes, there is often a class at car shows for drivers. Still, the basic classification is driver or show car; and you must decide which you want to have.

Concours—A concours, or show, car is the most expensive, and hardest to develop between the two. It is distinguished from the driver because it is seldom, if ever, driven. Generally it is trailered to a show and then trailered home again. This protects against many possible hazards to the car—fender benders, stone chips, mechanical failures and so on.

This car must look as good as or better than it did when it came off the assembly line. Every part must be brought back to original condition and quality. This is what the judging at the show is based on. For each incorrect part, points are removed. Depending on the judging organization,

either 100 or 500 points is normally the maximum given, with points removed in increments of a half or quarter. If you read national publications that advertise show cars for sale, you may have seen cars advertised as 98-point cars or even 100-point cars. This is what you would be striving for if you wanted a show car with which to compete in Mustang shows.

Obviously, the show car is going to be the most expensive—to buy, restore and maintain—of the the two classes. You can't hide or bury anything. And it's likely to bring the highest price at resale. The $20,000 someone paid for a 1965 convertible was probably because it was a show car with a number of trophies to its credit. However, the restorer may have had *more* than $20,000 invested in the car!

This is not necessarily wrong. You cannot put a price tag on fun or pride of workmanship. These are intangibles and part of the fun of the hobby. Here, too, is where budget and time come together. Maybe a $20,000 car is out of your budget for a

If you're not picky about originality, many "restified" cars, such as this clean '67 GTA with 428 Cobra Jet power, mags, traction bars and other mods, are enjoyable to own and drive.

Rough, but driveable and all there, this '66 convert was rescued from an Arizona Apache Indian reservation after having been torched for insurance purposes. Most all Mustang convertibles are worth restoring. Photo by Ron Sessions.

Basket case? Sure. But it's a '65 K-model convertible and all numbers match. This car *will* be resurrected.

Nice, original or restored Shelby, such as this '68, will bring two or three times the price of more common Mustangs. Photo by Ron Sessions.

one-shot deal. However, stretched out over a period of four or five years of building time, you might be able to afford it. Or, perhaps you haven't the patience to drag out a project for that many years. This must be part of your plan.

Driver—The second type of car is the one most often seen: the driver. As its name implies, it's driven frequently or every day. A driver is quick gratification and generally affordable because it may be bought in any condition. (Later I'll discuss condition as part of your budget.) Most restorers go for the driver because it is the least expensive,

can usually be driven while you work on it, and improvements can be made as you can afford them.

The parade car is a driver that has been nicely restored. These are the cars you see on display at the mall, at local charity car shows, driven in Fourth of July parades and turning heads at car-club runs. Drivers have generally been carefully restored to almost show quality in every respect. Usually, the undercarriage has not been painted and detailed as a true show car would be. The other details will be very similar.

Body Styles & Equipment—Now let's look at body styles as part of your budget. There are three body styles available: convertibles, notchbacks and fastbacks. Within these three body styles are further delineations—for six-cylinder and eight-cylinder engines, automatic and manual transmissions and special models such "K" cars, CJ's and Bosses. The combination of body style and engine and transmission will be one of the major determining factors of price.

Nice cars and still within the realm of affordability are '69—70 Mach Is. Watch for these to appreciate fast.

When in doubt, check the vehicle identification plate or sticker. This '69 suffered a quickie paint job and other abuse under past ownership. Photo by Ron Sessions.

The convertible is the most desirable model, with the V8 being the most desirable powerplant. Transmissions fall into the personal preference category. Some people love the feeling of control that a manual transmission provides—and four-speeds are much preferred over three-speeds. Others, brought up on automatic transmissions, think manual transmissions are primitive. I'll let you decide which is best.

Following the convertible is the fastback in preference. The notchback, however, runs a close third. The six-cylinder engine has its place where economy is concerned but is a distant second to the V8 in buyer preference.

A possible fourth class is the Shelby. This specialty fastback, or convertible, produced by the great race driver and marketeer, Carroll Shelby, is one of the most sought-after cars in the Mustang hobby. If this is your dream, be prepared to put a very severe dent in your budget, and only if you're able to find someone willing to part with theirs!

Then, of course, there's the other performance Mustangs—limited-production specials such as the Boss 302 and Boss 429, Mach I and Boss 351, and special-market oddballs such as the West German T-5, California Special and High Country Special.

Besides the Shelby, high-performance models and limited-production specials, the V8 convertible then, is the most desirable car, followed by the fastback and notchback V8 vehicles. This ranking of

desirability has many exceptions, but generally reflects the price you will have to pay for the car you want. Keep this well in mind as you plan.

HOW TO IDENTIFY

Before you run out and write that check for the Mustang of your dreams, you'd best know what it is you're looking at. The best way to do this is to read everything you can get your hands on that has information on Mustangs. A few books on the subject, magazine articles and knowledgeable friends can go a long way to helping you learn the basics before you jump in and buy. The more you know about the subject, the less likely you are to buy a car with hidden problems.

One of the first things to know is how to read a vehicle identification plate. It specifies the original color of the car, the engine it came with, the type and color of the interior trim and a host of other information about the manufacture of the car. This is important information. If the car has been altered in any way it will change the value.

Original Vs. Modified—Completely original cars are more valuable than cars that have been modified. Period. Likewise, a modified car will be cheaper for you to get into than will a low-mileage, totally original car. However, when you sell a modified car, even if you have spent great sums of money on it, you'll not get as much back as if that same amount had been spent on an original car.

Interior modifications are not as problematic as exterior or engine modifications.

You'll probably be redoing the interior of the car as part of the restoration. Look carefully, however, at panels that may have been cut out for speakers. Rear-quarter metal panels on early-model cars are hard to come by.

Look at the instrument panel. Sometimes, these have been cut to install various aftermarket products such as stereos, air-conditioners and other goodies. If you're going back to original, repairing an instrument panel can be very expensive.

An aftermarket sunroof is not too damaging to the interior of a car, but it will cost an arm and a leg to repair or replace the roof sheet metal.

Modifications to the body are the favorite tricks of the customizer and can create some serious problems. When large rear tires are installed, the car owner will often flare the fenders to allow clearance for the new tires. Check this very carefully. If the wheel openings have been radiused, repairs will be expensive, especially at the rear. So shy away from a car that's so modified. Also, you should be able to feel inside the rear wheel well and feel two layers of sheet metal spot-welded together along the radius of the wheel well. If you feel a rolled edge, it can be corrected by straightening.

Despite late-'60s-vintage wheels and Chevy front license plate, this '65 V8 notchback shows reasonable care and would make decent restoration candidate. It's much easier if the car is intact and running.

Holes cut in '69 rear-quarter trim panels for aftermarket speakers ruins these expensive and hard-to-find parts. Reproduction manufacturers are only just now beginning to make new parts for '69—70 Mustangs. Photo by Ron Sessions.

Aftermarket sunroof may have been somebody's good idea, but renders this '65 virtually worthless as a restoration candidate. Only proper fix is to cut off roof and weld on another—without the big hole. Photo by Ron Sessions.

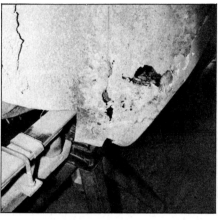

Three bad signs on this Mach I to beware of: aftermarket traction bars, rusted wheelhouse and plastic filler on wheel-opening lip. It pays to pull a wheel or two and have a look. Photo by Ron Sessions.

A '69—70 hood scoop had once adorned this '66. Now, hood must be replaced for restoration. Double-panel hoods do not respond well to the weld-and-plastic-fill method of repairing holes. Engine heat and vibration also work against this type of repair. Photo by Ron Sessions.

If trimmed however, the repair job will be much more difficult, but not nearly as bad as correcting a radius job.

Most other body modifications are readily apparent. Look for trim that has been removed and the holes filled. Look at the hood. Has an aftermarket scoop been added? Hood pins, maybe? Are the bumpers correct?

One of the really scary problems you may run into is someone selling a '65 or '66 fastback that has been modified to look like a Shelby. This is fairly easy to do and may pass to the untrained eye. Here, a working knowledge of the vehicle identification number will save you from a potentially serious problem.

Modifications to the engine include aftermarket speed equipment, bolt-on conveniences such as A/C or cruise control, and even complete engine swaps. Engine swaps are often done to remove a six-cylinder engine and replace it with a V8. If this has been done, it will quickly show up on the VIN.

Sometimes, a big-block V8 engine has been shoehorned in place of the original small-block V8—even on a '65—66. If this has been done, look for areas where the inner fender well or firewall may have been cut and sectioned to make room for the larger engine. Look also for Thunderbird and other big-block Ford engines used to

Hood pins, however nicely done, were not original on this '65 GT and are difficult to repair, short of hood replacement. Photo by Ron Sessions.

Problem with performance Mustangs is that younger owners tended to load them up with aftermarket equipment, tossing hard-to-find air cleaners, intake and exhaust manifolds, valve covers and smog equipment to the four winds. Photo by Ron Sessions.

However, some cars, such as this '69 Mach I 351, have been tastefully customized with Ford accessories.

History can really be divided into two great eras—Before Bondo (BB) and After Bondo (AB). Properly used, plastic body filler is the bodyman's best friend, but Bondo Abuse is rampant. Take a magnet when you go shopping for a used Mustang! Photo by Ron Sessions.

replace a Mustang big-block.

Corrosion Damage—Rust and plastic body filler are the next two big bugaboos to watch for when shopping for a car. Look carefully for rust. You might want to take a magnet along when you go looking at a car. Large areas of rust or plastic body filler will not attract the magnet. Look for rust in the front lower corner of the passenger door, in the front fenders at the bottom where the tire throws water and in the rear quarter panels just aft of the rocker panel. These are common rust areas.

Even more insidious is trunk-floor and rear "frame-rail" rust. The Mustang uses a *flange-mounted* fuel tank, wherein the fuel tank is dropped into a hole in the trunk floor and held in place with sheet-metal screws around its seam, or flange. Leaky weatherstripping lets water into the trunk area and eventually rust gains a foothold. In time, the floor or "frame" members rust to a point where the fuel tank can actually fall out of the vehicle—a definite safety hazard! Or the rear spring eyes and shackles will poke up through the trunk floor.

Also check the passenger-compartment floor pan for rust perforations—especially if your fancy runs toward convertibles. You can do this unobtrusively with an awl or Phillips-head screwdriver. Go around trying to punch the tool through the carpet. It will pass through the carpet and floor pan

Check for rust bubbles at lower edge of doors.

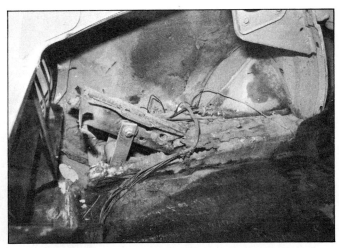
Trunk floor rust and rear "frame" rail rust can cause rear springs to crash through floor and fuel tank to fall out. Beware! Photo by Ron Sessions.

Likewise, check for floor-pan rust lurking under carpet. Take an ice pick or Phillips head screwdriver with you and try to poke through the sheet metal. Photo by Ron Sessions.

If the car has a vinyl top, look for telltale "bubbles" under the vinyl or a red-oxide staining of adjacent sheet metal that indicate roof rust. Photo by Ron Sessions.

If possible, take the car to a shop and have a look underneath. This '69 is sagging because left rear spring of car above came up through a rusted floor. Photo by Ron Sessions.

if there's a rust spot. If you get down to the serious business of taking the car for a ride, stop by your favorite mechanic who has a hoist, raise the car and then check the underside for rust.

You will have to weigh these modifications and problems against your budget. Any modification will incur expense to bring it back to original. If, of course, the modification is something you like and want, go for it. There is nothing that says you must only have an original car. For example, I really enjoy a good stereo system. My own car has been modified to accommodate a good system and I love it. And note that good-quality stereo systems are now available to fit the existing radio and speaker cutouts *without* modification. So you *can* have your cake and eat it too! Go in with your eyes open and you'll come out with your pocketbook still intact.

WHERE TO FIND A CAR

Finding a Mustang is not a really big problem. Ford produced nearly 3 million of its spunky ponycars up through 1970. And a great number of these are still on the road. It's not as if you were looking for an old three-wheel Messerschmidt or perhaps a GTO Ferrari. There are still many Mustangs to choose from.

Begin your search in the local newspaper. This will give you an idea of what

Begin with a visual inspection of the car. Do the body panels align? Is the side of the car wavy when you sight down alongside?

Rear shackle kit and beslicked chrome reverse wheels should throw up the yellow flag of caution. Photo by Ron Sessions.

costs are in your area and what is available. Next, tell everyone you know that you're in the market for a particular year and model of car. Be prepared to hear, "Oh, I wish you'd have told me last week. My sister just sold her '65 convertible with 20,000 original miles for $150!" Unfortunately, it happens and you'll never find a deal as good. So be forwarned that it will happen—after you make your purchase!

When telling everyone you're in the market, be sure to tell the people who can best help you: the folks at the gas station you frequent, the people at the parts house, repair shop and other places catering to the automotive crowd.

I seldom recommend used-car lots. Usually, you will be wasting time. If and when they get a good one in, it goes to a friend. If they put a dog on the lot that the friend doesn't want, it's generally overpriced. They know what they have and won't be giving any good deals.

Some communities have a special weekly newspaper dedicated to selling used cars. These are often excellent places to locate a car, especially in the larger communities. Often good deals may be found through these "auto trader" or similar magazines. The secret is to find out what day they arrive at the newsstand and be there to get your copy first. In this case, the early bird catches the Mustang.

There are a number of national publications such as *Hemmings, Collector Car News, Mustang Monthly, Fabulous Mustangs, Mustang Illustrated, Super Ford,* and others that can be of great help in your search. Often, this is the only way for rust-belt dwellers to find a rust-free car from the Southwest. A few hundred dollars for a plane ticket could be much less expensive than some major bodywork to replace a floor pan. If you choose this way to go, be sure you get detailed photographs of the car, and a qualified appraisal before setting off to buy.

One of the best ways to buy a car is as a member a Mustang club—one with a local chapter nearby is best. Here, you'll find folks who are walking data banks on the subject of Mustangs. Not only what is correct and what is wrong on a specific car, but where good cars are located. Don't expect, however, to find a car the first night you go to a meeting. Mustang club members take their membership very seriously. You'll be welcomed when they see you too are serious about membership and car ownership. After a time, the good cars will begin to appear.

Most club members will be very straightforward with you. They want you to be successful and don't want to get a reputation within the club of taking advantage of fellow club members. So, join a club, whether you're looking for a car or already have a car. It can help you and you can help it, thereby keeping the hobby moving forward.

HOW TO BUY

After you have found the Mustang of your dreams, there are still a few things to do before breaking out your wallet. At a time like this, emotion can often get in the way of good sense. So, to maintain equilibrium, take a friend along as a disinterested third party. You'll be way ahead if your friend is a qualified mechanic or really knows Mustangs.

Begin with a visual inspection of the car, taking into consideration all of the things mentioned above. Then, check out the electrical components, lights, horn, radio, heater, A/C, and the rest. Start the car and listen for knocks, shakes, rattles and rolls. Move the car a few feet to test the brakes and transmission.

Go outside the car and push down hard on one of the fenders and quickly release. The car should rise up and quickly and not bounce up and down if the shock absorbers are good. If tire wear is uneven or if you hear any untoward groans from the undercarriage, have the chassis checked out professionally. While checking the suspension, look at the car from head on. Does it list to one side or the other, front or rear? Again, this could point to suspension problems if the car does list.

Next, check under the hood. With the car running, *carefully* remove the radiator cap and check for exhaust bubbles coming from the coolant, a sure sign of a blown head gasket, cracked head or block. Check the

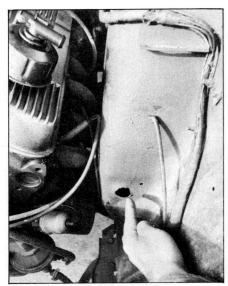

Check under the hood, too. Previous mechanic cut holes in inner fenders to lube upper control-arm pivots—a travesty all too common and very expensive to fix. Photo by Ron Sessions.

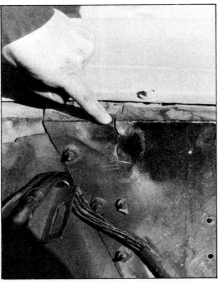

Corrosion at inner-fender bolt flange caused this ominous stress crack. Panel should be replaced. Photo by Ron Sessions.

oil dipstick. Emulsified water in the oil is also a sign of a crack in the head or block. Dirty, gritty, oil will indicate less than regular maintenance. Does the engine compartment look well-maintained? Does the owner care about it?

Take the car out for a test drive. How does it accelerate, brake, feel and run? Relax your grip on the steering wheel. Does it pull one way or the other? Apply the brakes with the same relaxed grip on the wheel. Does it pull right or left? If the car passes all of these tests, stop in for a prearranged visit to your local mechanic and have him do a compression test, leak-down test and possibly a smog check. If after all of this, the car still looks good, it's time to make an offer.

In today's market, almost everyone overprices the car because they expect to be talked down in price. It's just the way car business seems to be done—even on new cars. You wouldn't haggle over a television set at Sears, but it's expected when you buy a car. Begin by making an offer about 30% lower than the asking price. Expect a counter offer about 10% lower than the asking price. Somewhere between 20% and 25% below the asking price is generally a reasonable area. Having shopped for a long while, you will have a pretty good idea of what the "right" price is for the car.

Remember to use positive psychology when negotiating. Say that $900 is all you can afford rather than tell the seller his car is only worth $900. All this will get you is an angry seller; exactly what you don't want.

Keep these things in mind as you go looking for "your" Mustang. Know what you can afford and what you want. Buy the best Mustang you can for your money.

Take along a friend to check the car out. Maybe he'll stumble into or *through* something you might have overlooked. No, this is not Fred Flintstone but Don Kott of Tucson Frame Service. Photo by Ron Sessions.

3

Disassembly

Before you begin tearing your car apart, it's wise to do a little prior planning to prevent possible problems. Start with the area in which you will be disassembling the car. A complete restoration can tie up a two-car garage for the duration of the project. You may want to consider renting a place to store parts that won't be worked on for awhile. This will provide more room in the garage and more work space for your project.

For additional storage space, add a few shelves, clear some of the unused junk out of the rafters, and reorganize the rest of the work area. I make a bunch of wire hooks from coat hangers that can be hooked over the rafters to hang various small parts.

Position the car so that both doors may be opened with room to pass between the door and any obstruction. The car will also be jacked up about 24 in. off the floor, so this clearance should be maintained to that height.

PREPARATION

Let's take a look at some of the tools and materials you need to make the job safe, fun and easy. I'll assume you have a fairly complete set of hand tools or can borrow them. Where needed, many specialized tools can be rented from a tool-rental store. These include such items as a cherry picker for lifting the engine, a Porta-Power for heavy bodywork, a Morgan-Nokker for less-heavy bodywork, and other small, but not often used, hand and power tools.

You'll also need a set of jack, or safety, stands and a hydraulic floor jack for the duration of the project, so the option of renting these would be costly. Buy the best you can afford. Your life depends on them. I prefer jack stands of heavy-gage construction, with flat-plate bottoms and forged contact pads. When it comes to floor jacks, stick with a name brand that the local hydraulic repair service carries parts for, and go no lower than 1-1/2-ton capacity. Stay away from swap-meet, fixed-caster, stamped-steel cheapies from across the Pacific!

As you disassemble the car, you'll wind up with thousands of parts. You must

Dave Cawthorne's '65 convertible before it underwent complete restoration. Entire body was "dipped" to chemically remove all paint and rust.

organize them in some fashion in order to get them all back together into a running car. Be sure you have a wide selection of cardboard boxes, coffee cans, heavy-duty plastic sacks (better than paper bags because you can see parts to identify them), masking tape to label with and a *large* number of shop towels. There is no such thing as too many shop towels!

I strongly suggest having a Polaroid camera at hand. Use it to take pictures of parts and areas before you disassemble them. This will be a wonderful help when it comes time to put it all back together. Finally, keep a pencil and pad handy to make notes, inventory parts, make plans and generally keep yourself organized.

SAFETY

You're probably as tired of reading about safety as I am writing about it. However, even as I write this, someone is being injured who either didn't know, read, or follow the safety rules. Let's look at them again and promise yourself never to break them.

Never get under a car unless it is supported by safety stands. This means never jack up a corner of a car and stick your head in between the tire and fender. If the jack were to slip, blow a seal or in some way move, its bye-bye head! Always use a safety stand even if you're only working on one corner.

Disconnect the battery when doing any electrical work. This will prevent a possible short circuit that could damage the alternator or any electronic components or cause a fire. I remove all jewelry, including my wedding ring when I'm working around

REWIRING YOUR MUSTANG
by Ron Sessions

Getting the job done—day in and day out—is the electrical system in your Mustang. Yard upon yard of wire snakes through your Mustang's innards, connecting the far reaches of that sexy sheet metal with the starting, charging, ignition, lighting and accessory circuits. The original electrical system installed in your car at the factory goes about its tasks unobstrusively, year after year. Rock solid, with no moving parts. But after 20 years or more, things can and do happen.

Maybe it's a wire that slowly chafes through its insulation, shorts to ground and, poof, you've got one brain-dead Mustang on your hands. Or perhaps its a corroded terminal that won't let even one more electron Boogaloo on through after years of exposure to corrosive salt spray. Packrats and mice have been known to gnaw away insulation and wire to line their nests with—so it pays to start the engine and move the car once in a while just to keep those little critters guessin'.

An engine compartment is a particularly severe environment for automotive wire, exposed as it is to temperature extremes, vibration, oil and road debris. And if you're bringing a basket case back to life—a victim of a serious collision, engine fire, or previous "race-car" modifications as performed by the Chimpanzee Brothers—rewiring your car can become labor-intensive.

Sure, frayed or cracked insulation can be temporarily bolstered with electrical tape. And damaged sections of wire can be snipped out and new sections spliced in using soldered butt connectors and covered with shrink wrap. But really, the only *right* way to repair damaged wiring is to replace it—either individually or by complete harness—terminal to terminal. And if a harness meltdown occured because of a circuit overload or short circuit, better find the cause before laying new wire in there—or you may be in for a repeat performance.

When replacing individual wires, keep wire gage, connector type, insulation type and insulation color code in mind. You probably know that most of the wire used in Mustangs is PVC (polyvinylchloride) wrapped and in the 10 to 18-gage range. The *higher* the gage number, the *smaller* the diameter of the wire (and the fewer the strands). So, a 10-gage wire can handle big amp loads with less resistance than a 12-gage wire, and so on. High-amp-draw components include: power-seat and power-window motors, headlamps, starter solenoid, cigar lighter, and air-conditioner, to

Frayed insulation in rear lighting harness is a short waiting to happen. Photo by Ron Sessions.

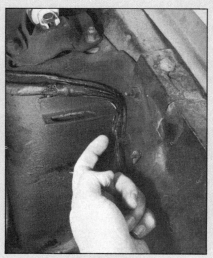

Engine compartment is tough environment for wiring—heat, oil, vibration and harsh chemicals have attacked this harness, leaving exposed wire. Photo by Ron Sessions.

Best way to repair wiring is by replacing complete harness, page 130. Photo by Ron Sessions.

name a few. Circuits that draw less current can get away with higher (numerical) gage wire—items such as a clock, back-up lamps, power door locks, instrument-panel lighting and the ignition system.

If in doubt as to the correct wire gage to use, make sure the replacement wire is the same or *lower* (numerical) gage as the original. Where the original wire may have been sheathed to protect it from high heat or chafing, make sure the replacement has the same protection. And, if at all possible, try to maintain the original wire color coding with the new wire. That means replacing a 14-gage yellow-with-green-stripe wire with the same article. It may look neat to replace your engine compartment wiring with all blue wires until the first time you have to troubleshoot an electrical problem and no longer have the color coding to guide you. Then, you'll be reduced to pulling and tugging on one end to see what moves down in the harness—no fun!

Pay attention to routing. Keep wires away from hot exhaust manifolds or sharp edges that may chafe away at the insulation. Use cable-ties to keep the wires organized into bunches. Whenever a wire passes through a bulkhead, use a rubber grommet to protect the wire insulation. If you have access to one, follow the wiring diagrams in the factory shop manual or reproduction literature such as that offered by Jim Osborn Reproductions, Lawrenceville, GA 30245. And work with only *one* wire at a time to help minimize confusion.

The best way to go, especially if your Mustang is undergoing restoration, is to replace the entire harness. Original factory harnesses are available for late-model cars, and for the earlier models, any number of reproduction harnesses. Before ripping out the old harness, it's a good idea to make small "flags" out of masking tape to identify mating connectors. Once the old harness is out of the car, lay it down next to the new or repro harness to make sure the two match. Reproduction-harness quality varies, and you may have some work to do to make the new harness work. The best method is to buy a high-quality harness in the first place. Ask club members or other restorers how they faired with the harnesses they've bought. If you plan on replacing more than one harness, do them one at a time.

Before removing hood, draw outline of hinges on hood to facilitate installing in same position. Then, trace outline of hinges on inner fenders in same manner.

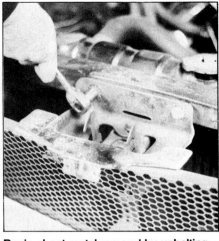

Begin sheet-metal removal by unbolting hood-latch assembly.

any possible hot wires. The metal can cause a short between two wires and give you a terrible burn.

Keep flames and sparks away from the battery. A battery gives off hydrogen gas, even when not charging. The introduction of a spark or flame to this gas can cause an explosion. Remember the *Hindenberg* dirigible? I've seen this happen. A friend, who should have known better, got in a hurry and did some welding next to the battery. When the battery blew, he got acid in his face, acid on the car's new paint and started a fire in the engine compartment.

Never suspend a block and tackle from the garage rafters to use as an engine hoist. You're just liable to find the garage roof on top of your Mustang. Although strong enough to collectively support the roof, most individual garage joists or rafters are not strong enough to support the weight of an engine.

Always use the right tool for the job. This means use a box wrench or socket when breaking a bolt loose, not an adjustable wrench. Never use a screwdriver as a pry bar. Rent or buy the right tool. You will no doubt save money in the long run.

Keep your work area clean and free of clutter. Leaving parts and tools scattered about is a keen way to loose tools and parts; and an equally keen way to trip and fall. Be especially cautious of spilling oils or solvents on the floor. Wipe them up *immediately*. I buy kitty litter (the cheapest kind) to use as an absorbent for spills. Pour it on, let it absorb the spill, then sweep it up. You can also buy a commercial absorbent, but kitty litter is one-half to one-third

the price and works about as well. Keeping these things in mind, let's begin taking this car apart.

FRONT EXTERIOR TRIM & SHEET-METAL REMOVAL

Bumpers, Valance & Grille—Begin front-end removal with the bumper. The easiest way to remove the bumper is to remove the bumper brackets from the frame, removing the brackets and bumper as a unit. Just be careful to support the bumper and brackets to avoid scratching any sheet metal. Don't forget to remove the bolt attaching the bumper to the fender at each end of the bumper. On '65—68 models, remove the upper stone deflector.

The grille on each year is a little different, but the process remains basically the same. After removing the hood lock assembly, unscrew the upper and lower grille-flange retaining screws and lift out the grille. On '69—70 models, remove the three snap-in plastic rivets from the top and one screw from the hood lock assembly.

If you're working on a GT, remove the wires from the fog lights. Finish the nose of the car by removing the grille reveal molding, the headlight doors, headlights and headlight buckets. If these fasteners are rusted in place, drench them with penetrating oil and remove with an impact driver.

Next, remove the valance panel, removing the bolts and disconnecting the wires for the parking lights, where applicable.

Side-marker lights are used on '68—70 models. Disconnect their leads at the quick disconnect.

Remove the horn(s) next. If so equipped,

Next off are headlight doors and headlight assemblies.

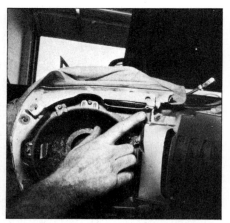

On '65—66 models, headlight buckets tend to crack here. Repair is by replacement. Photo by Ron Sessions.

It's a bit of a tight fit when removing these grille bolts. A stubby ratchet makes things easier.

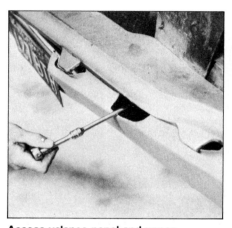

Access valance-panel and upper stone-deflector bolts through this gap between the two.

If so equipped, rocker-panel molding must come off to remove fenders. There's one stud nut at each end and several of these clips riveted to the rocker panel.

Doors—The doors are a snap to remove. If you have not already done so, be sure to disconnect any wiring running through the door and door pillar. Now remove the four hinge-to-body bolts. Support the door on a floor jack with a 2x4 on the pad and have a friend stabilize the door. It's not a one-man operation! If, for some reason, you just want to remove a door without first removing the fenders, remove the hinge-to-door bolts first.

Rear-Quarter Windows—Remove the upper adjusting stop. Remove the clips from the regulator rollers. Lower the regulator and remove the four window-channel bolts. Move the channel out of the way. Slide the glass out, then the channel.

Exterior Trim—Two exterior trim pieces should be noted here: the die-cast chrome rear-quarter side sculpture and Shelby/Mach I fiberglass brake-cooling scoops. Your Mustang may have either, but not both. In any case, each is removed after removing three stud nuts from inside the car behind the rear-quarter trim panel, beneath the rear-quarter window.

The remainder of the exterior trim may be pried off gently with a flat-blade screwdriver or pushed out from behind the sheet metal.

REAR EXTERIOR TRIM & SHEET-METAL REMOVAL

At the rear of the car, begin disassembly with the trunk lid. The only problem, outside of simple hinge removal, is to remove the spring bracket that fastens the deck spring to the lid on the fastback.

The taillights and rear-quarter caps are

remove the A/C condenser. When disconnecting its fittings, use a flare-nut wrench and be careful not to bend or break off the condenser elbow.

Next, remove the rocker-panel reveal moldings. These are retained with stud nuts at each end, accessed from inside the front and rear wheel wells. After removing these nuts, slide each molding off the remaining trim clips. These clips are riveted to the body the full length of the rocker panels, and can be removed, if necessary, by drilling out with a 1/8-in. drill bit.

Front Fenders—Here, things get a little bit more complicated because there are several hidden bolts. The first is located in front of the door post just above the door hinge. This is illustrated on the following page. The next one is behind the kick panel. You may have to remove a rubber plug, after removing the panel, to access the bolt. Finally, there are two bolts in the wheel well that attach the headlight housing to the fender.

Now, remove the bolt at the bottom rear of the fender and the six fender-to-engine compartment bolts. *Gently* lift the fender up and outboard of the wheel and set aside. And note the number and placement of any fender-to-body shims.

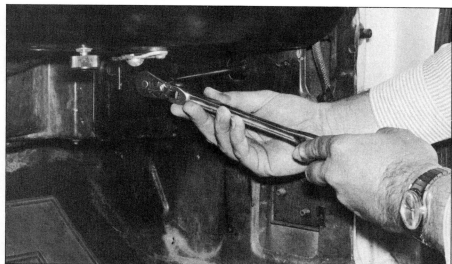

These two bolts are well hidden and must be removed before the fender will lift off the car.

If door sags or detents are worn out, be sure to use new door hinges at assembly time.

With fenders off, make sure this drain hole isn't clogged with debris. Clogging here causes cowl-vent rust and a resulting water leak into the front footwells. Also inspect condition of rear splash guard and rubber seal (arrow). Replace if necessary. Photo by Ron Sessions.

secured with nuts inside the quarter panels. Remove the rear valance and then the bumper. Rear side-marker lights are removed by undoing the nuts from the retaining studs and disconnecting the leads. Keep each lamp and retainer together.

The fuel tank presents only one problem: early-model tanks are equipped with a drain plug while later-model tanks are not. Tanks without a drain plug should have the gasoline siphoned out or run completely dry. WARNING: *do not attempt to siphon gasoline by inserting a hose down the filler tube and sucking with your mouth!* Leaded gas does great things for pre-'71 engines, but is poisonous and can be fatal if ingested or inhaled. Instead, use a siphon pump and practice safety precautions while performing this operation.

You can, of course, remove a fuel tank full of fuel, but remember that gasoline weighs almost 7 lb per gallon—so 10 gal of gas will increase the weight of the tank you're removing by 70 lb! And that weight may shift from side to side as the gas sloshes around inside! So, get as much gas out of the tank as possible.

Mustangs use a flange-mounted, or drop-in-type, fuel tank wherein the top of the tank forms part of the trunk floor. To

With door supported, remove six door-hinge-to-body bolts and lift off door.

Mustang fuel tank is flange-mounted to trunk floor with sheet-metal screws. Pull up underlayment to access them.

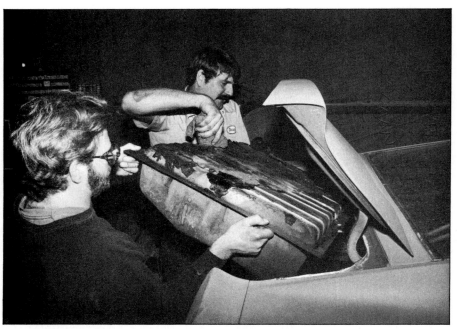

To remove tank, drain as much fuel as possible and enlist helper to assist lifting out bulky tank through small decklid opening.

ENGINE & DRIVE-TRAIN REMOVAL

Preparation—Although I've worked on cars all my life, I never liked getting greasy all over. Therefore, I keep things as clean as possible. You'll like it too, and the job will go easier. If the car is still mobile, start by steam-cleaning or high-pressure water/solvent-spraying the engine compartment and undercarriage before putting it up on jack stands.

Most communities have a car wash with a high-pressure water/solvent-spray bay. If not, some commercial garages will steam-clean your Mustang for just a few dollars. If you're going to water/solvent-spray your car yourself, you'll need a couple of cans of engine degreaser, two large plastic Baggies (big enough to cover the carburetor(s) and distributor), masking tape and some old clothes and shoes.

Before you leave for the car wash, warm-up the engine for about five minutes. Saturate the really greasy parts of the engine with engine degreaser. By the time you get to the car wash, the grease will be loosened and ready to come off. Before you start rinsing off the degreaser with water, wrap the distributor and carburetor in Baggies, holding them in place with masking tape. Keeping water and condensation out of the distributor and combustion chamber will make the car much easier to start afterward. Put your old shoes and clothes on and go to work.

If no steam-cleaning or high-pressure-water-spray facility is available, you can do the job using a degreaser only. For light cleanup, commercially available spray cleaners such as Gunk are satisfactory. The cheapest way to do it is to buy a gallon of degreasing compound and mix it with a solvent according to directions. Then, apply this mixture with the family garden sprayer. *Warning:* Do not use gasoline as a solvent. Use kerosene or regular automotive solvent. Gasoline and lacquer thinner are too volatile to use.

Working in an area where it will not damage plants, the street, or offend the neighbors, warm the engine to operating temperature and thoroughly spray the engine, engine compartment, transmission, undercarriage and rear end of the car. Of course, you will have wrapped the distributor and the carburetor as described for steam-cleaning the engine. Let the degreaser soak in for 10—15 minutes. Then, using as strong a stream of water from the garden hose as possible, wash off *everything* you degreased. The degreaser acts as an *emulsifier*—lifts and keeps in

remove the tank, loosen the hose clamps, disconnect the filler tube from the tank and remove the filler tube from the body by removing its retaining screws.

Disconnect the fuel line and plug it to prevent gas from leaking out. Disconnect the fuel-gage sending unit and then remove the bolts securing the tank to the floor pan. Lift out the tank. Finally, remove the rear wiring harness.

suspension—and lets the grease become water-soluble. Whether by steam-cleaning or chemical degreasing, you now have a clean engine and drive train. Let's get things together and start getting the engine out.

Tools & Equipment—In addition to your hand tools, safety stands and a hydraulic floor jack, you'll need the following: masking tape; pen or pencil; assorted cans and boxes for small nuts, bolts and parts; a drain pan for engine fluids; a can of penetrating oil and a roll of mechanic's wire. If you don't have a creeper, a heavy piece of cardboard or a piece of Masonite will make crawling under the car much more comfortable. Remember to get out the Polaroid camera and take pictures of such things as wiring connections, throttle linkage, clutch linkage and anything else you're liable to forget. Now you can begin.

HPBooks has thoughtfully provided you with several books on the section I'm about to cover. With the exception of six-cylinder Mustangs, HPBooks offers a detailed removal and installation section in each of three titles: *How To Rebuild Your Small-Block Ford*, by Tom Monroe; *How To Rebuild Your Ford V8 (351C, 429)*, by Tom Monroe; and *How To Rebuild Your Big-Block Ford*, by Steve Christ. Because it has been covered so well in these books, I'll only hit the high points and the things you should especially look out for.

Battery—Begin by removing the battery. It's in the way of many components you'll be working on, is a source of corrosion and explosive gas, can send sparks flying if metal tools dropped thereon go from the positive terminal to ground, and may become discharged in time.

Loosen the two clamps and *rotate*, rather than pry, them off the terminals. Prying a cable end off a terminal with a screwdriver may easily pull a terminal out of the battery—for good. Set the battery aside and keep it charged up using a 1—6-amp trickle charger. Discharged batteries tend to *sulphate*—form deposits on plates—which eventually turns them into so much lead scrap.

Hood—Hood removal is a two-man job. Enlist the help of a friend to prevent dropping and damaging the hood or scratching and dinging the fenders or cowl. First label and disconnect any leads for hood-mounted turn-signal indicators or engine-compartment courtesy lights. Then use folded rags under the rear corners of the hood to prevent scratches, and remove the bolts and hood.

Normally, you would mark hood loca-

Having drained all fluids, Dave begins removing radiator.

tion with a marker or scribe to get it back in the right position. You're going to remove the hinges and paint the car, however, so you'll have to carefully realign the hood when the restoration project is done.

Where Do I Start?—Begin by draining the fluids from the radiator and oil pan. This will prevent a big mess on the floor. In the accompanying photos, good friend, Dave Cawthorne, has removed all front-end sheet metal before removing the engine. On a restoration project, this is the prefered order. The sheet metal must come off, and this way, you do not have to lean over a fender to access the engine. But, if you aren't going to remove the front sheet metal, feel free to do it any way you wish. There is no right or wrong way.

Radiator—Remove the radiator hoses, both top and bottom. If they do not remove easily by gently twisting them back and forth, simply cut them off. High-mileage hoses should be replaced anyway. Don't risk damaging the radiator by wrestling with them.

Disconnect the heater hoses now, as well. Next remove the automatic-transmission cooler lines, if so equipped. On these and any other lines, such as the fuel lines, use a flare-nut wrench to prevent rounding the corners of the nuts. Run a hose between the two cooler lines or cap them to prevent automatic-transmission fluid (ATF) from dribbling onto the floor.

Unbolt the fan shroud and lay it forward over the fan. Next, remove the four bolts

holding the radiator to the radiator support and pull it out. Store the radiator in an area where the fins will not get bent or damaged. Now you can remove the fan.

Carburetor & Linkage—Of course you will remove the air cleaner first. Don't forget the hot-air duct if your car is so equipped. Here is where your Polaroid camera will come in handy. Take a couple of shots of the carburetor linkage and routing of the fuel and vacuum lines. This will be very helpful when it's time for reassembly. For added insurance, put masking-tape labels on particularly confusing connections. Remove the linkage and lines.

Hoses & Lines—Mark and disconnect all hoses and lines from the top and sides of the engine. If you plan a complete restoration, remove everything you can, mark it and keep it with its associated parts. Keep vacuum lines with the brake booster, and fuel lines with the carburetor. Double-check your work. If you lift the engine with some hose or wire still connected, chances are they will break.

Accessories—If your car is equipped with an air conditioner, you must bleed off the freon before disconnecting any refrigerant lines or removing the A/C compressor. When discharging freon from the system, *always wear safety glasses and gloves*. Freon is a hazardous substance. In the compressor, it is in a liquid state. When released, it comes out as a gas under extreme pressure, way below freezing. Severe frostbite injuries can result from freon "burns," including blindness.

To bleed off freon, put on your glasses and gloves and slip a flare-nut wrench onto the connector. Next, wrap a rag around the connector to deflect freon as it bleeds off. Carefully loosen the hose connection until you hear a slight hissing noise, then stop. This is the freon escaping. Allow the freon to bleed off *slowly*. After the hissing sound has stopped, finish disconnecting the line. Immediately tape off all open ends on the hose and compressor. This helps prevent contaminating the system with dirt or moisture. Now you can finish removing the A/C compressor, followed by the power-steering pump.

Further Top-Side Work—Over on the right side of the engine, remove the air pump and associated plumbing and hardware, if so equipped. Air pumps were installed in 1968-and-later 49-states cars and in '66-and-later Mustangs sold originally in California. This exhaust emission-control device was called a Thermactor.

Also remove the alternator or generator. Label and disconnect the wires, loosen the

On late-model cars with Drag Pac, disconnect and remove engine oil cooler.

Disconnecting exhaust on big-block Mustangs is no picnic due to tight quarters.

Big-block 428CJ leaves little room between exhaust manifolds and shock towers for unbolting engine mounts. But somebody has to do it.

adjusting bolt, slide the alternator or generator toward the engine, remove the drive belt and unbolt and remove the alternator or generator.

Clutch & Shift Linkage—On cars with manual transmission, remove the clutch linkage. Remove the spring from the equalizer bar to the firewall and the equalizer-to-release bearing arm; then disconnect the pushrod extending through the firewall. Remove the equalizer-bar bracket and the equalizer. Do this by undoing the bolts on the outer equalizer bracket, the one closest to the fender. Then the bracket end equalizer will lift out. From inside the car, remove the shift lever from its bracket by removing the two bolts in the top center. You do not have to disconnect manual-transmission linkage from under the car.

On cars with an automatic transmission, disconnect the manual-lever control rod from beneath the car. The selector lever can be removed by removing the dial housing, selector-lever plate, then the housing-and-lever assembly.

Final Top-Side Work—Double-check that all disconnections have been made between the engine and chassis. Look again at the fuel line, throttle linkage, wiring harness at the coil and distributor, heater and radiator hoses, plus the starter wiring. At the back of the engine, remove the engine-to-firewall ground strap.

With everything finished on the top side, jack up the car and set it on jack stands. You can place stands under the crossmember below the engine, on the frame where the crossmember bolts up, or use the box-section sheet-metal "frame rails" farther

back. Do not use the suspension to support the car. It is unsafe, and you'll be removing the suspension later anyway.

Manual Transmission—The transmission sits on a crossmember, secured by two bolts extending through this crossmember mount. You must remove the two nuts holding the transmission to the mount. Next, disconnect the speedometer cable at the transmission. This is easy to forget and as the transmission moves forward with the engine, the cable gets permanently kinked and must then be replaced. Disconnect the emergency-brake cable from the equalizer lever. If so equipped, remove the backup-light switch.

Automatic Transmission—If, in an unusual case, you wish to save your ATF, work according to the following directions: Remove the cover plate on the converter housing, then remove the two converter drain plugs and drain the fluid into a *clean* pan. Then, reinstall the two drain plugs. The drain plugs are visible behind the bell-housing dust shield, recessed slightly into the driveplate. You'll need to rotate the crankshaft to bring the drain plugs into view.

Next, loosen the transmission oil-pan bolts and drain the fluid at one corner of the oil pan. When all of the fluid has drained, tighten the cover.

Of course, draining ATF for reuse is rarely done because new fluid is necessary unless it was just recently added. If you wish, you could drain the ATF now to avoid excessive leakage out the rear of the transmission during removal.

For transmission removal, you must dis-

connect the transmission-cooler lines from the case. It's also best to disconnect the filler tube. Next, remove the manual and kickdown linkage rods or cables from the control levers. Remove the neutral-start-switch wires from their retaining clamps and connectors. Then, double-check to see that everything is disconnected.

Drive Shaft—Now, move to the rear of the car where the drive shaft joins the rear axle. Match-mark the drive shaft to the rear axle with paint or center punch. Then, unbolt the drive shaft at the rear universal joint and slide it rearward off of the splined transmission output shaft.

When you begin lifting the engine and transmission, some gear oil or ATF will spill out the rear of the transmission. Prevent this by plugging the opening in the extension housing with a plastic plug made for this purpose or a small paper cup. If you have nothing else, use a shop towel and masking tape.

Exhaust—Unbolt the header pipes(s) at the exhaust manifold(s) and allow them to hang down. For extra clearance underneath, it's a good idea to remove the exhaust system completely.

Mounts—There still remains the matter of the engine mounts and transmission mount. Support under the engine with a jack and unbolt these from the chassis. Also unbolt the transmission crossmember from the

Lifting Engine & Transmission—You're now ready to pull out the engine and transmission together. Check everything over once more. Take the car down from the safety stands and let it rest on the floor. Position a floor jack under the transmission. If it is an automatic transmission, place a wide board between the jack and the transmission pan to prevent damaging the pan. Attach a lifting plate to the carburetor mounting base, or bolt a chain diagonally across the engine. Use thick, large-area washers so the bolt heads cannot slip through the chain links. Accessory mounting holes in the cylinder-head ends make good attachment points.

Now slowly begin raising the engine. Immediately, jack up the transmission until the two mounting bolts on its case clear the crossmember. Give the engine a little more lift and let it slide forward. With a little coaxing, the engine and transmission will then lift out of the engine compartment. Voila!

Strip Engine Compartment—With the engine and transmission out of the car, finish removing everything you can from the engine compartment. Take off the master cylinder and brake line across the firewall. Remove the brake junction block from left inner fender. When disconnecting fluid lines such as fuel and brake lines, problems may arise unless you are very careful. Use a good penetrating oil on the fittings. This will loosen them a bit. Next, use a flare-nut wrench or wrenches to finish loosening the flare-nuts. Use of these wrenches, as seen in the illustration, prevent rounding the corners of the nuts.

On cars with disc brakes, disconnect and remove the proportioning valve. Remove the accelerator rod or cable and the clutch rod (manual transmission only). Accelerator cables are disconnected at the pedal inside the car, then pushed through the firewall after the grommet has been worked free. Accelerator rods need undoing at the pedal also. Then they can be unbolted from the firewall.

Remove the front wiring harness that crosses over between the two headlights, marking all connections. Mark the main-harness wires at the voltage regulator, remove the wires and then the regulator. Disconnect the wires and remove the stoplight switch. Remove the starter's remote relay and finally, remove the blower motor.

Leave the steering box in place for now. That will be dealt with after you remove the steering column. Everything but the steer-

Cherry picker is in place and ready to lift engine. With front fenders removed, reaching things in engine compartment is a little easier. Ease engine up a little at a time, making sure nothing hangs up. Notice homemade, extra-volume oil pan.

Remove all interior trim around windshield and backlite.

In most restorations, battery tray is corroded from acid and should be replaced. Soak bolts with penetrating oil beforehand. Photo by Ron Sessions.

ing box should now be out of the engine compartment and you can turn your attention to the interior.

INTERIOR REMOVAL

Windshield & Backlite—Although the windshield and backlite, or rear window, are not generally thought of as part of the interior, removing them makes removal of the interior much easier—particularly the headliner and instrument-panel dash pad. Our demonstration car is a convertible, but there are many similarities between the convertible and a notchback or fastback.

Begin by removing all trim from the windshield and backlite. This includes the exterior stainless-steel moldings as well as the interior garnish moldings and any window louvers, like the bolt-on Boss 302 units. On the convertible, the header molding is removed by accessing four nuts through the interior trim molding as seen in the photo above. The rest of the moldings and garnish come off after removing the many Phillips-head screws.

To remove the exterior windshield or backlite reveal molding on a notchback or fastback, a special tool, K-D 2038 or

equivalent, is required. To use the tool, force one end of the hooked blade under the molding and find each retaining clip. Pry the clip away from the outside edge of the molding. With the hook on the tool, release the catch between clip and molding. After releasing all of the clips, the molding will easily lift out. There are four sections of exterior molding around the windshield and backlite of the notchback and fastback. On the convertible, the lower windshield trim molding can be left in place until the glass is removed. It will be reinserted into the butyl-rubber seal when the window is replaced.

1965—68 Models—With all moldings removed, cut away the interior lip of the

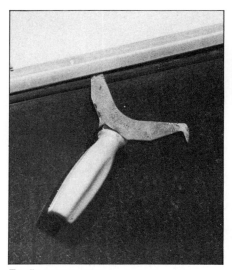

To disengage exterior reveal moldings from clips, slide hooked end of tool under molding and pry each clip away from outside edge.

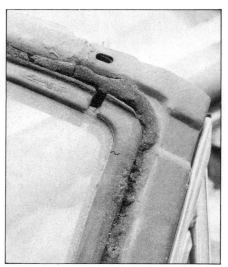

On '65—68 Mustangs, windshield and backlite mount in weatherstrip-type seal.

Cutting off lip releases windshield from its frame. When pressure is applied to push glass out, there is less chance of breaking it. Photo by Michael Lutfy, courtesy of Mustang Illustrated magazine.

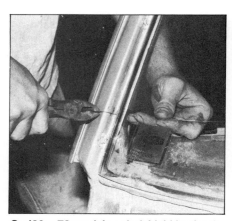

On '69—70 models, windshield is glued in place with butyl rubber adhesive. To remove, cut seal with seal cutter or piano wire. Photo by Ron Sessions.

Push gently and evenly in the top center.

rubber windshield seal with a razor or sharp mat knife. Again, this serves for the backlite as well. Now, by gently pushing from the inside of the windshield or backlite, near the top, it will push right out. Have a helper ready to grab it. Do not pry on the corners of the windshield or backlite with a screwdriver. Doing so may crack a lami-nated windshield or shatter a tempered backlite. Make sure the gasket seal is cut all the way through around the full circumfer-ence of the glass.

1969—70 Models—Beginning in 1969, the windshield and fastback backlite are "glued" in place with butyl rubber adhe-sive. The *only* way to remove one of these is to *cut* it out with a special windshield-seal-cutting tool shown on this and follow-ing pages or with piano wire strung be-tween two stout wooden dowels.

Steering Wheel—Removing the steering wheel now eases access to the instrument panel. Begin by removing the horn ring. To do this on '65—67 models, push in at the

Here are tools you'll need to remove '69—70 Mustang windshield and fastback backlite. Photo by Ron Sessions.

Push in and rotate *counterclockwise* to release horn ring on '65—67 models.

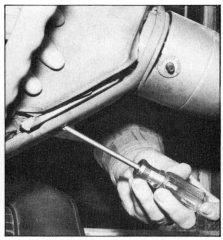

Starting in '68, horn ring or pad is retained to wheel with screws.

center of the horn ring, or button. Rotate it about 1/8-turn counterclockwise. This releases the horn ring, or button, for removal. Don't lose the spring. 1968-and-later models have a bar-type horn ring that doubles as a crash pad. Go around to the backside of the steering wheel to remove the two or three screws. Once the screws are out, the horn ring will lift off.

Next, unscrew the steering-wheel retaining nut. Usually, a couple of real strong tugs will break the wheel loose. Just be careful on '67-and-later models with collapsible steering columns to avoid pounding on the wheel. If this dosen't work, use a steering-wheel puller. Now, with the steering wheel out of the way and the windshield removed, you have easy access to the instrument panel.

Disassemble Instrument Panel—You're going to be removing a lot of wires from here on. I suggest you mark *every* wire and connection. This will save huge headaches at reassembly time!

On models so equipped, remove the Rally-Pac from the steering column by loosening the screws below the steering column. On '66 models, there's a wire guide that must be removed from the steering column and instrument panel. Disconnect the wires and remove the unit. On '67 Shelbys, remove the underdash gage pod. Later Shelbys place these gages in the console, so you'll have to deal with them when removing the console.

Often, steering wheel will pop loose by pulling rim while pushing from behind with your knees. If this fails, use a steering-wheel puller. Photo by Ron Sessions.

On early models with Rally-Pac, unbolt clamp from column, disconnect wires and remove.

Underdash A/C Unit—On 1965—66 models, the optional air-conditioning system is a simple hang-on unit. Some '67 models without factory air had a similar hang-on unit that was dealer-installed. Regardless, it blocks access to underdash components and the console, so take it out first.

Before you begin, however, *the unit must be discharged of all freon!* Go back and read the section on discharging the air-conditioning system, page 39, before you remove the evaporator.

When you're sure that all freon has been bled off, disconnect the hoses and wire to the evaporator, remove the four retaining screws and drop the unit from underneath the dash.

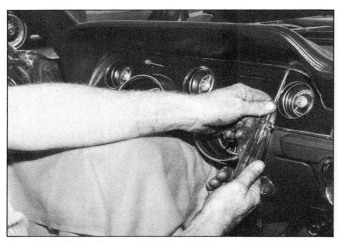

Cover steering column with shop cloth to prevent scratches as you remove instrument cluster.

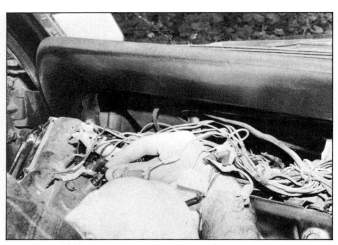

On '65—66 models, be sure to label instrument-panel-wire connections. You'll be glad you did at assembly time, even if you plan to replace the harness.

On '67—68 models, hidden nut for cluster is above ashtray. Remove ashtray screws and disconnect cigar-lighter lead. Then reach up and unscrew sheet-metal nut.

Instrument Cluster—Place a thick towel over the steering column to protect it from scratches. Reach up behind the cluster and loosen the speedometer-cable retaining nut or plastic clip and disconnect the speedometer cable from the speedometer. You might want to forgo this step and simply remove the cluster from the instrument panel. The cluster will come out of the panel slightly, allowing you to reach between the cluster and the panel and undo the speedometer cable. The trouble is, not all cars have enough slack in the cable to let the cluster come out far enough for you to reach the speedometer cable. Then you have no choice but to stand on your head and remove the cable from behind the instrument panel. Also, on '69—70 cars, the cable is retained by a plastic lever. All you have to do is press down on the tang and pull the cable free.

1965—66 Models—Up through '66, instrument-cluster removal is super simple. Remove the perimeter Phillips-head trim screws, pull the cluster from the instrument panel, disconnect the speedometer cable and lift the cluster out. Well, there is one fly in the ointment. All the wiring to the instruments must be removed at each instrument. By all means, mark the wiring as you disconnect it. The cluster is marked with wire colors, but that's not really a good substitute for making your own identifying tags.

1967—68 Models—The '67—68 clusters are more involved. Here, you must first remove the ashtray. Pull the tray to its open position, push down on the tang and remove the tray from the instrument panel. Now use a stubby Phillips screwdriver to get at the three screws holding the ashtray inner panel to the instrument panel. Push the panel back and to one side and disconnect the lead for the cigar lighter. Once this inner panel is out, you can reach through the hole to get at the instrument-cluster retaining nut hiding at the upper right. An alternative is to remove the radio, and get at the instrument-cluster retaining nut through the radio's larger panel hole.

As you go parts-hunting, you'll see a lot of '67—68 clusters with the right edge broken off. Now you know why.

Once this hidden nut is removed, the instrument cluster can be pulled away from the instrument panel. If you haven't disconnected the speedo cable, do so now. You'll also see a lot of wiring running all over the back of the instrument cluster. Look for the multi-plug connector at the left end of the cluster. Disconnect it and the single wire running to the windshield-wiper switch. There may also be another single wire running to the tachometer, if your car has one. Once you have these wires disconnected, the cluster will come free.

1969—70 Models—Yet another procedure is needed for '69—70 cars. The dash pad must come off first, then the instrument cluster. To remove the dash pad, start at the outer, lower corners. At each corner you'll find an end cap that curves around the end of the instrument panel. They are retained by two screws, one on the side and one at the lowest corner. Remove both screws, and then the end cap. Under the end cap is the lower dash-pad retaining screw. Remove it. You'll also find three retaining screws at the dash top, two in the

Also less than obvious on '67—68 models is need to remove screws for heater/A/C controls to free cluster.

On '67—70 models, pull cluster out, reach behind and disconnect two multiple-plug connectors and speedometer cable.

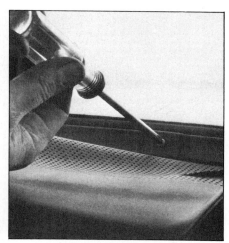

On '67—68 models, dash pad is retained by sheet-metal molding at base of windshield.

center section where the pad dips down between the two "cowlings" and three screws each under the driver's and passenger's cowlings. Take all of these screws out and lift off the dash pad.

Now you can turn to the instrument cluster. Look for a screw under both the speedometer bezel and tachometer or fuel-gage bezel. Another two screws are at the top of the cluster. Remove these four screws and the cluster will be loose from the instrument panel. There is still the speedometer cable, multi-pin connector and single wire to the tachometer to deal with. You can either reach over the top and get through the hole opened by the missing dash pad, or pull the cluster out some and get them that way.

Glovebox—On all models, turn to the glovebox next. To remove the glovebox, open the door and remove the retaining cable (on early models,). From under the glovebox door, remove the door retaining screws and lift the door and glovebox liner out as a unit. Once the glovebox is out, undo the glovebox courtesy light and set it aside.

Instrument-Panel Dash Pad—If you are working with a '69—70 car, you've already removed the crash, or dash pad, so skip ahead to the next heading. All you '67—68 owners will have to wait until the rest of the panel is apart to remove the dash pad; the cluster, glovebox and instrument-panel trim pieces must come off before going for the dash pad. It's the trim pieces that delay dash-pad removal on this body-

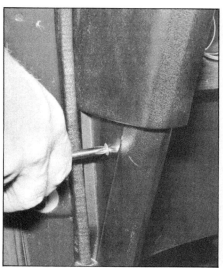

On '69—70 models, don't forget to remove dash-pad screw at either end of instrument panel.

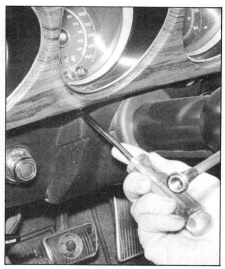

With dash pad removed on '69—70 models, cluster removal is a simple matter of taking out a few screws.

style. The trim pieces are retained with some Tinnerman sheet-metal nuts threaded onto studs formed on the trim pieces. It takes all the room you can get from removing the radio, glovebox and so on to get at these nuts. You must use a wrench on them, so don't try to shortcut here. It will only end up costing you some skinned knuckles.

On 1965—66 models, remove the chrome moldings and retaining screws from the underside of the dash pad. Also remove the molding at the windshield-dash-pad intersection. Remove the radio-speaker grille and speaker. This is the only way to access two dash-pad retaining nuts beneath the speaker. Remove these two nuts. With a pair of long-nose pliers, re-

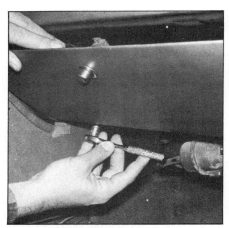

Removing glovebox frees up space to get at other under-dash components.

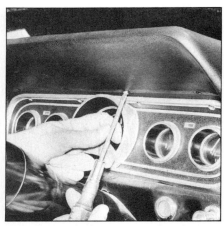

Start dash-pad removal on '65—66 models by removing these screws along the pad's underside.

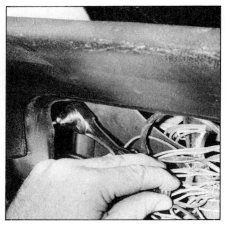

On '66 models only, don't forget to remove nuts at lower corners of instrument panel.

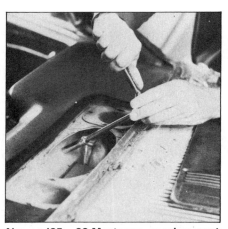

Also on '65—66 Mustangs, speaker must come out to access center mounting studs for dash pad. Defroster ducts are next to remove.

Next off is radio and, on '67—68 models with roll-top forward console, roll-top frame.

move the defroster-duct retaining clips and then the defroster ducts. On 1966 models, there are two retaining bolts at each end of the pad. Remove the nuts from these bolts before attempting to remove the pad. Be extremely careful removing the pad if you plan to use it over again. It tends to stick to the instrument panel and will crack if not handled with care. If necessary, free it from the instrument panel by wedging a putty knife in between.

On 1967—68 models, remove the heater-control assembly. With that out of the way, you can reach through the opening and remove the retaining nut at the left end of the panel. In the middle of the panel, remove the radio. Pull off the knobs and bezels, undo the retaining screws and the radio is free.

Disconnect the antenna, power leads and other wiring. The ground wire is separate and requires only a slight loosening of the retaining bolt so the spade connector will slide free. It is easiest to reach the ground bolt through the glovebox opening.

If you have a console, the radio and console come out separately. But you might as well take the console out now. Remove the radio, watching for radio-to-console hardware inside the console compartment—the one with the roll-up door. The console itself comes out in two parts. First is the upper trim piece, then the console itself. Look for trim-panel hardware in the rear ashtray.

After the trim panel is off, take off the

To remove automatic-transmission shift lever, unscrew Allen-type setscrew and lift off T-handle.

And on '67—68 models, lift out console upper half.

If equipped with console, lift out rear ashtray and remove console retaining screws.

Remove screws retaining shift plate to lower console half.

piece around the shifter. It requires four screws to be removed. Finally, the console-to-drive-shaft-tunnel screws can be removed and the console lifted out.

OK, now that the radio and glovebox are out of the way, take off the end of dash ventilation outlets. Look for screws holding the rubber ducting to the metal outlets. Remove the screw and pull off the outlets. The instrument panel should be bare except for the trim pieces and dash pad.

Roll over so you can reach up under the dash and look for the sheet-metal nuts that retain the trim pieces to the instrument panel. Get your wrenches out and turn these nuts off. They are tough customers because they like to cut their own threads on the way off. Once you've got them off, the trim pieces will be free. Lifting them off exposes the dash pad, which curls around the two sheet-metal support "eyebrows" to anchor below the trim pieces. As long as the windshield-to-dash-pad molding is off, the dash pad should pull free.

On '69—70 models, you should already have your panel down to the basics. All that remain are the switches and so on. These are covered in the next heading. You may still have the section above the glovebox in place; it removes with screws. Look around the edges for any hardware. With the dash pad out of the way, it's not to difficult to see how everything goes together.

You may have a console in place. If so, remove it now. Look for hardware in the glovebox portion of the console, at its rear end. Read the section on console removal for the other years to get the general idea for console removal. But, beware of one '69—70 requirement. The passenger seat must be removed. If it isn't, you'll probably break the console trying to twist it around the high seat backs. Don't risk breaking this expensive and difficult-to-find part; remove the seat.

If you have a Shelby, disconnect the gages at the front end of the console after you have moved the console back a little.

Another tip is for all those people who love to take everything apart because, well,

it just needs to come out, doesn't it? Leave the heater, and especially the heater/air conditioner alone in the later chassis. Removing these parts is the great torture of all time for Mustang mechanics, so don't do it unless there has been a fire or flood. These underdash parts are well-protected anyway, and taking them apart usually results in more damage than help.

Heater Control, Switches & Knobs— The rest of the removal operation is fairly simple and straightforward. There are a

With all attaching hardware removed, dash pad can be lifted (or pried) off sheet-metal eyebrows and removed.

After removing screws retaining console to driveshaft tunnel and disconnecting any wiring, console can be removed.

few operations, however, that should be explained.

On '65—68 models, disconnect the three heater-control cables—temperature-valve, heater-air-door, and defroster-air-door—from the heater. On '69—70 models, there are but two disconnections: temperature-control cable and air-door-control cable.

Pull the control unit free of the dash and label and disconnect all electrical leads and vacuum hoses. Then, remove the heater-control assembly and cables as a unit.

To remove the light switch, pull the light-switch shaft into the "on" position. On the bottom of the switch housing is a small spring-loaded button. Press in on this button and pull the light-switch knob and shaft out of the switch housing. On '67—70 models, you must first remove the two bolts for the parking-brake control and drop down the lever to gain access to this button from below. A bezel tool, page 49, will remove the retaining bezel, and the switch housing can be removed. Use this same tool on the wiper switch after removing the knob. In a pinch, you can use a small flat-blade screwdriver, inserted into one of the bezel notches 90° to the bezel face, and tap the bezel loose with a hammer or the butt of your hand. Be careful to avoid slipping off the bezel notch and scratching the instrument-panel face.

The dash-mounted ignition switch used in '65—68 models is easy to remove. First, place the key into the switch. Then, insert a paper clip or length of mechanic's wire into the small hole directly to the right and below the key. By rotating the key toward the accessory position and pulling out on the key, the tumbler can be removed. Remove the bezel, unplug the ignition-switch leads and the ignition switch can be removed from the instrument panel.

Finish instrument-panel disassembly by removing the air-vent knob and emergency brake as a unit. Remove the choke handle, if so equipped. Finally, remove the cigarette lighter (if you didn't do it already). This is done by first removing the wire from the back. Next, with the lighter element removed from the base, insert two fingers into the base and hold it rigid. From the back, unscrew the socket with your other hand.

Wiring Harness—Disconnect the remaining wires at these locations: heater motor, courtesy-light switches, (remove kick panels and switches from door pillar and disconnect wires), turn signal and turn-signal flasher, floor-mounted dimmer switch, stoplight switch and fuse block. You may wish to leave the wires attached to the fuse block and remove the block itself.

Now, all main-harness wires have been disconnected. Reach through the opening in the instrument cluster and disconnect the harness from the harness clamps along the underside of the instrument panel at the firewall. Check for any other connections. If there are none, begin to draw the wiring harness out through the opening for the

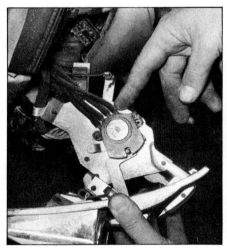

Label heater/A/C control wires, cables and vacuum hoses to ease assembly later on.

instrument cluster. Things will tend to hang up as they pass throught the firewall. Help them along by going to the engine compartment and push them through.

At this point, the instrument panel should be completely bare. If you plan on chemically stripping the paint by "dipping" the body, remove the windshield-wiper

After removing light-switch knob, unscrew bezel with screwdriver.

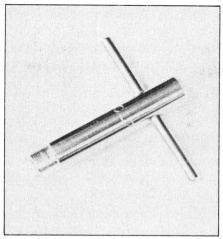

Tool is for removing bezels.

Dash-mounted ignition switch on '65—68 models comes out by inserting key, turning switch to ACC position, inserting paper clip in hole shown and pulling on key.

motor and wiper transmission. Otherwise, leave these in place if the wiper motor was working properly. Other items to remain if you wish are the clutch- and brake-pedal assemblies and steering column. I recommend, however, they be removed for repair and reconditioning if you desire a ground-up restoration.

Heater Core—There is plenty of room now to access the heater core.

On '65—66 models, if you didn't remove them as described previously, remove the control cables. Cut the heater hoses off the core with a razor or mat knife. Wedging a screwdriver down between a stuck hose and core fitting is a good way to ruin the fitting. Besides, you should replace the hoses anyway. If the heater motor hasn't been removed, remove it now along with the ground-wire-to-instrument-panel retaining screw. Disconnect the nuts that retain the heater and remove it.

On '67-and-later models equipped with air conditioning, the heater core is integrated into the A/C evaporator assembly. If your Mustang has this system, jump ahead to the heading *"Combined A/C/Heater Assembly."*

Steering Column—There are only a few steps remaining to pull the steering column at this time. Remove the retaining screws from the steering-column weatherseal on the firewall and remove the seal and cover plates. Disconnect the steering column-to-instrument-panel clamp and remove the steering column. Remove the retaining

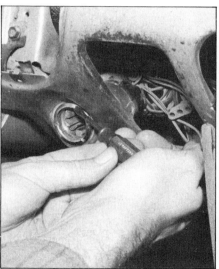

Bezel can be tapped loose with small flat-blade screwdriver as shown. Then, disconnect switch and remove.

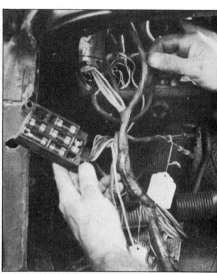

Dave is removing fusebox along with main wiring harness.

screws from the steering column weatherseal on the firewall and remove the seal and cover plates. Unscrew and remove the turn-signal lever to prevent damaging it during steering-column removal. Remove the steering column.

Combined A/C/Heater Assembly—If you absolutely insist on removing the combined air-conditioner/heater assembly used in '67-and-later models, here goes. First, disconnect the hoses and refrigerant lines. Then, remove the vacuum supply tank. At

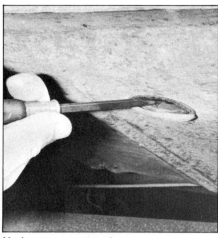

Heater assembly on all '65—66 cars and '67—68 models without factory A/C is prone to water damage at cardboard plenum (arrow). Plastic replacements are available from aftermarket sources.

Under car, remove plugs to access front-seat retaining nuts.

the engine-compartment side of the fire-wall, there are two evaporator-case mount-ing stud nuts and one blower-housing stud nut. Remove these nuts. Back inside the car, reach through the glovebox opening and disconnect the vacuum hoses from the reheat and outside-recirculating vacuum motors.

Disconnect the control cable from the temperature-blend door and the wiring from the air-conditioner thermostat switch. Next, remove the evaporator rear-support bracket screw and the blower-housing support-to-cowl-panel screw. Now, you can move the blower housing over. Pull the rubber drain hose up and out of the floor pan. Push the evaporator case back and down; then you can disconnect the follow-ing wiring and hoses: the vacuum hoses from the heat and defroster motors and the wiring from the air-conditioner clutch switch, blower-motor resistor and the blower motor itself. Now the blower hous-ing and the A/C defrost plenum-chamber assembly can be removed from the car by sliding it over to the right and out.

The remainder of the parts are removed in the following order: water-valve vacuum switch, the two retaining screws, and then the switch. Next, the upper frame and temperature-blend door screws, the shaft and lever assembly and the retaining clip. Now you can remove the lower frame by removing the four screws retaining it. By removing the retaining clip on the reheat-door lever and moving the motor arm out of the way, you can remove the shaft and lever assembly and the reheat door. All that re-

mains is to remove the heater core.

Final Details—The instrument panel is clean now. Check for any little items you may have missed. Did you remove the an-tenna? How about the defroster-duct hoses? Get everything out because you're going to paint the instrument panel along with the rest of the car.

You still have all the interior trim to remove. Begin by removing the front and rear seats.

Seat Removal—To remove the front seats, remove the four rubber plugs in the floor pan under the car. Remove the retain-ing nuts with a 1/2-in. deep-well socket. Remove the two rearward nuts first, then the front nuts. If you remove the front nuts first, the seat can fall backward binding the rear nuts.

The rear cushion is removed by pushing back at the base of the cushion about 1-1/2 in. until the two hooks are released, then lifting it out. Notchback or convertible seat backs will come out after removing the two 7/16-in.-hex sheet-metal screws at the bot-tom. Push the back up, releasing it from its two retainers just below the package tray. Remove folding-back fastback seats by un-bolting pivot brackets at floor. Fixed fast-back seat backs are unbolted at floor and top right corner.

Front Window Cranks, Door-Latch Handles & Armrests—All model cars have the door handle and front-window crank retained with a Phillips head machine screw unless the door handle is of the retracting variety. Remove this screw and the door-latch handle or window crank can

then be removed. On '68—70 models, a thin brushed-aluminum plate was stuck over the top of the screw hole. Carefully remove this by running a thin-blade knife under the edge. Be careful not to bend it as it will be glued back on when reused.

In 1968, a folding, retractable door han-dle was introduced. this is removed after the door panel is off the car. Two bolts retain it to the inner door.

The deluxe door panel on the '67 slips over the door handle. You'll have to re-move the panel, then the door-latch handle.

Unless the door panel has a molded armrest as with the deluxe '67—70 trim, the armrest is retained by two sheet-metal screws. Use a 3/8-in. nut driver to remove these screws. In 1968, Ford inserted a plas-tic plug in the screw hole. This may be removed with a flat-blade screwdriver.

Door Trim Panels—On '65—68 Mus-tangs, the door panels are retained to the door with spring clips. To release these clips, insert a flat-blade screwdriver di-rectly under the clip and pry out. Be sure you get under the clip rather than under the fiber board. Any pressure on the fiber board and it will crack.

For '69—70, a full door panel with an upper garnish molding was introduced. The garnish molding is retained by two Phillips-head screws, one in each corner. Remove these, remove the door-latch but-ton and rotate the garnish up and out. The rest of the door panel is removed as de-scribed above. Don't forget to disconnect the wires to the door courtesy light, if so equipped.

Using 1/2-in. deep-well hex socket, remove four nuts and lift out each front seat.

Put these front-seat carpet protectors in a safe place to reuse later.

Lift up front of rear-seat cushion and unbolt center seat belts. On fastbacks, belts stay with cushion, as shown.

If your car is equipped with a door-pull handle, remove the two covers by popping them out with a screwdriver. Then remove the two retaining screws.

Remote-control-mirror actuators are removed by taking out the two retaining screws or removing the bezel with a bezel remover tool. If you cannot find one of these, use a small screwdriver and hammer—just as with the headlamp-switch bezel. Set the blade of the screwdriver in the notch in the bezel. Tap on the handle with a small hammer. This should rotate the bezel, loosening it. I try to avoid this method as much as possible. There is a real good chance of scratching or breaking the bezel.

The remote-control mirror comes off by removing the unit from the door panel and disconnecting the cables from the control.

To remove the power-window switch, loosen the door panel and pull it forward. Disconnect the switch-assembly wiring at the block connector. When the door panel is out of the car, remove the two retaining screws and retainer plate from the switch and lift it from the panel.

Kick Panels & Rear-Quarter-Trim Panels—Kick panels are easily removed. Remove the screws retaining the panel to the cowl then slide the panel out. Removing the rear-quarter-trim panels is a similar operation, except that screw location is not as obvious.

Remove the window crank as you did for the door panel. Note that on some of these rear-window cranks, they are retained to the shaft by a small offset Allen-head screw

Sheet-metal screws hold bottom corners of seat backs in place.

as shown in the photo, page 52. Remove the center screw for the quarter-trim panel next to the window crank. There is a screw at the bottom front edge, one at the top rear edge and one midway down the rear edge. When these are removed, the panel can be taken from the car.

Roof Console—On '67—68 models so equipped, remove the two screws at the front of the console. This will allow it to drop. Unclip the two rear retainers. Disconnect the wires and the console is out.

Headliner—The headliner should be removed after the windshield, backlite and door-seal weatherstrip have been removed.

On early '65 ('64-1/2) Mustangs only, door-latch handles and window cranks are fastened with retaining clips, *not* with screws as on later models. Use a GM-type door-handle remover tool to disengage clips, as shown. Photo by Michael Lutfy, courtesy of *Mustang Illustrated* magazine.

Pull the vinyl trim off around the door opening. Pull all of the glued edges away around the doors and window frames. At the rear quarter, the bottom of the headliner is retained by a cat-claw. These are sharp hooks that hold the headliner in place. Lift the headliner off these hooks. At the package tray, lift the metal tabs, release the

51

On '65—68 Mustangs, simply pry door trim panels loose with screwdriver. Get screwdriver blade directly under each spring clip to avoid breaking fragile fiberboard panels.

On some models, rear-window cranks are retained with offset Allen-head setscrews.

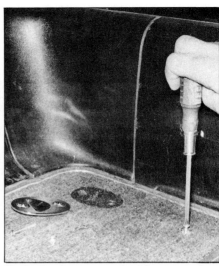

On fastback, rear-quarter trim removal begins with rear flooring.

Take out fixed rear section, then undo hex bolts retaining folding section and lift out.

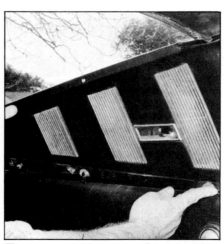

Then, remove trim screws for extractor vent trim pieces.

fiberboard retainer and pull the headliner out.

Working from the front, remove each top bow in order. Pull down on the bow in the center, rotate it forward and remove it from its retainer holes. There are two holes here, side-by-side. Mark the hole your bow came from so that it goes back in the same hole. On the last bow, there are two bow hooks. These hold the back bow in place. Slide them off the bow and put them away for safe keeping. Remove the bows from the headliner, numbering them 1 through 4 from front to rear.

Floor Console—If you haven't already done so, remove the floor console next.

On '65—66 models, there were actually two types of consoles used: one with the under-dash air-conditioner, the other without. Without air-conditioning, the console had a glovebox in the front. To remove the console, first remove the gearshift knob or handle. Remove the three screws from each side of the console at the floorpan. If you have a glovebox, there are two screws to be removed in here. Lift the console from the rear and rotate it out.

On '67—68 models, first remove the gear selector-lever handle. With a flat-blade screwdriver, gently pry the top-cover pad loose and remove it. From behind the glovebox door, remove the two retaining screws from the shift-lever opening-cover and remove it. Going through the glovebox opening, remove the two screws that hold the radio to the rear support bracket. Remove the three screws from each side of the console at the floorpan and slide the assembly back. Disconnect the antenna lead and associated wires from the radio. Disconnect the console lamp wire and remove the console from the car.

On '69—70 models, work through the glovebox door and remove the retaining screws and the top finish panel from the console. Remove the screws that hold the console to the shift assembly and to the instrument panel. The rest is removed as above.

Carpet—The carpet is the last to go. Remove the left and right sill plates. Remove the bolts the hold the seat belts to the floor-

Find all quarter-trim-panel attaching hardware and lift out panel. Now you can access outboard rear seat belts.

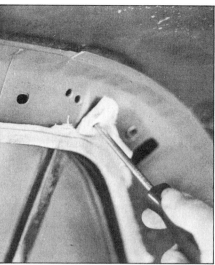

To remove headliner, first pull off vinyl trim from door opening. Unscrew coat hangers, visor mounts and A-pillar trim pieces.

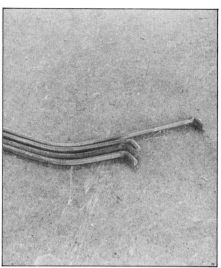

Unhook header bows from slots in roof and drop headliner down. Headliner goes into dumpster, but keep bows in order for reuse in new headliner. Number from front to back with tape.

pan. Remove the screws from the plate around the gearshift lever.

Where the front and rear carpets meet, just outboard of the front seats, there is a screw holding the carpet to the floorpan. If you can't find it, lift out the carpet and you'll see where it's held down. Remove these screws. Some models also had screws in the floorpan near the firewall. As you lift the carpet out, you'll run into them. They must also be removed.

By this point, you should have a pretty barren automobile. It will be even more barren after you remove some of the sheet metal.

UNDERCARRIAGE

In this last section, we finish disassembling the rest of the car except the front and rear suspension. Believe me, it's best to keep the car on its wheels as long as possible so that it is mobile. We'll save the disassembly of these components for later.

Once again, jack up the car and support it on jack stands, allowing yourself room to crawl under the car. Finish removing the fuel lines. Next, remove each of the brake lines going to each wheel, and finally, the emergency-brake cable.

Carefully check the flexible brake lines. Bend them into a tight "U" shape and look for cracks. Do this all along the line, especially at the connectors. Because your life is at stake here, its best to replace these

With rear-quarter trim removed, reach into rear-quarter opening on notchbacks and convertibles and remove quarter window and channel. Side-sculpture-trim retaining nuts can then be accessed.

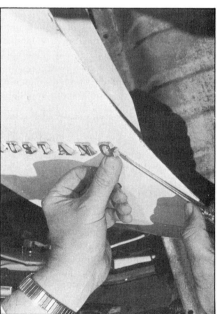

Most Mustang script can be carefully pried off from outside or pushed out from behind.

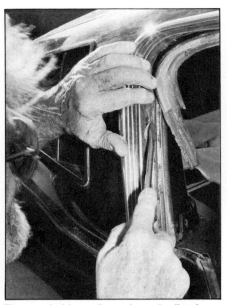

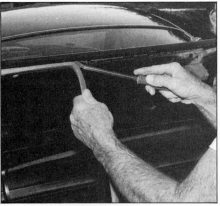

Soft trim and seals such as weatherstripping should *always* be replaced in any restoration. Start door-opening weatherstrip removal by undoing screw at bottom of windshield pillar. Once screw is removed, weatherstrip can be pulled out of channel.

Pieces of old weatherstrip and adhesive remaining in channel should be scraped out prior to installing new weatherstripping. A flat blade screwdriver works fine.

Decklid weatherstripping is glued in place. Remove by working stout screwdriver or narrow gasket scraper under adhesive bond while pulling down on loose end of weatherstrip. Remaining adhesive can be removed with lacquer thinner or scraped off with screwdriver after heating area with torch adjusted for low heat.

hoses if there is any doubt about them.

All that's left is the exhaust system. If this is to be completely replaced, use an oxyacetylene torch or an exhaust-pipe cutter to pull components apart. If you wish to save any components, you'll have to carefully disassemble them. When it comes to separating unions between pipes or between pipes and muffler, heating them just below red hot will help to free these joints.

Congratulations! You've just completely torn your car apart. I know there is a sinking feeling in the pit of your stomach about now; but worry not, we'll get it all back together better than new.

Some repair shops and specialty service shops offer professional steam cleaning. It's worth the money.

4

Suspension

When you drove your Mustang home, did it shimmey around 45 mph? Perhaps it tended to float up and down as you traveled along. Each time you hit a bump did it sound like the front end was going to fall out of the car? If so, you can bet your car has suspension problems and probably some steering problems, too.

Even if your car seems to handle well, there are parts that need attention. It is just part of working with a 20-year-old car. It's also part of a restoration to rebuild the car to like-new condition. Therefore, you need to remove the entire suspension and evaluate it piece by piece. Count on replacing all rubber parts, all ball joints and other links.

This section details how to correct suspension problems by completely rebuilding the suspension, or by repairing just the the parts that show obvious problems. Besides being necessary if you want to call your car "restored," it is easier and cheaper in the long run to completely go through the suspension in one step.

Steering is covered in the next chapter, but is an integral part of the front suspension. Therefore, read the steering chapter before tearing into the front suspension. Perform all tests given there so you know what to expect during disassembly.

When you have rebuilt the suspension, it will give you like-new ride and handling. Furthermore, if you want to take advantage of modern technology, you can make your daily driver easily outperform the original. More on the latest technology later.

CHECK FOR WEAR

Shock Absorbers—Shock absorbers are usually the first suspension components to fail. These usually have a life span of 20,000—50,000 miles, depending on the quality of the product and type of service.

To determine if the shocks should be replaced, push down hard on either the right or left front fender and quickly release it. The bouncing movement should end abruptly. If the car continues to bounce or float up and down, the shock is worn out and should be replaced. Perform the same test on the remaining three corners.

Front-suspension kit from Kanter includes all parts needed to rebuild Mustang front suspension. These lower ball joints bolt to lower control-arm assemblies, but some suppliers sell ball joints *riveted* to new lower control arms. The choice is yours. Photo courtesy Kanter Auto Products.

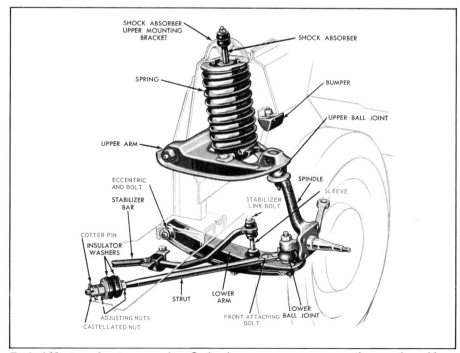

Typical Mustang front suspension: Spring is on upper arm, as was the practice with early Ford unitized bodies. Drawing courtesy Ford Motor Company.

Ball Joints—Check the condition of the front suspension upper and lower ball joints. Using a floor jack under one of the front "frame rails," raise the car until the tire is off the ground. Have a helper rock the wheel and tire while grasping it at top and bottom.

While he does this, watch for play. There may be some play from the wheel bearings, as shown by the wheel rocking independently of the steering knuckle. This can be corrected by torquing the nut at the outer end of the spindle to about 20 ft-lb. To access this nut, remove the wheel-bearing dust cap.

Rocked more vigorously, the suspension will show upper and lower ball-joint wear. Watch the intersection of the steering knuckle and the lower and upper A-arms. If the steering knuckle moves and the A-arms don't, the ball joints are worn. The upper joint is the critical one on Mustangs. Because the spring seats on the upper A-arm, the upper ball joint is the one that takes the weight of the car. Expect it to show wear first.

In fact, Mustang upper ball joints are notorious breakers. Needless to say, this could really ruin your day if it breaks while you're tooling down the interstate. The wheel and tire will fold under the car, rendering all steering, braking and accelerating commands useless. Don't be too surprised if you find the upper joint flopping around like a wet Mackrel. It's a common fault on high-mileage Mustangs.

Upper A-arm Inner Pivots—Another early Mustang problem is worn and squeaky upper A-arm inner pivots. Instead of installing a zirk fitting at these points, Ford merely put a plug in the grease hole. Thus, at every chassis lube, the mechanic was supposed to remove the plug, and apply grease with a needle-tipped grease gun. Needless to say, not too many early upper A-arm pivots got lubed. The result is worn, squeaky pivots. Push up and down on the front fender as in a shock-absorber test and you'll here these joints squawking away. Fixing these noisy joints means a complete front-suspension disassembly.

These are the basic front-suspension tests. If you are doing a complete restoration, rebuilding the entire front end should be considered part of your restoration project. The Mustang rear suspension is not as critical as the front. Other than testing the rear shocks for wear, you only need to check the rear springs for loss of arch—*sag*.

Rear Springs—To check the rear springs—leaf-type springs—select a spot

Remove both nuts from bottom of shock absorber to free it from upper arm.

on each side of the body at the rear from which to get a ground-to-body measurement. The top edge of each taillight bezel or top of each rear-wheel opening is convenient. Measure the distance from these points to the ground and note them. Give the rear of the car a sharp push downward, then slowly release it and let the car come to rest untouched.

Remeasure from the points selected to the ground at each side of the car. If there's more than 1/2-in. difference between the two measurements, suspect that the springs may be fatigued and should be re-arched.

Because this isn't a positive check for sagged springs, continue checking. Step back and look at the car from the side. If it looks low at the rear—leaf springs are more likely to sag than the coil springs at the front suspension—you can bet the rear springs have sagged. As a final check, look underneath the rear of the car. If the *camber*, or arch, of the springs have reversed—bowed up in the center at the axle pad instead of being flat or bowed down as they came out of the factory—they've sagged and should be re-arched. This should have shown up as excessive bottoming when the car was driven over mild bumps.

Leaf-spring re-arching is not a do-it-yourself operation. To have this done, remove the springs from the car and take them to a qualified spring shop. During the

re-arching process, they should also replace the rubber bushings in the spring eyes and insert insulators between the spring leaves to prevent squeaks. This job is best done with a press, mandrel and other necessary tools, but it can be done with hand tools.

Once you've completed all suspension checks, you should be ready to rebuild the suspension. Start by disassembling the front suspension.

REMOVE FRONT SUSPENSION

Remove Shocks—Start by removing the front shock absorbers. Undo the top shock mount while the car is sitting on the front tires. This will compress the springs so the shocks won't be loaded in their fully extented positions.

Once you've unbolted each upper shock mount from its bracket, remove the bracket. Push on the end of the piston rod to compress the shock, moving the upper mount down through its bracket. Once both sides are done, raise the car and place it firmly on jack stands so all four wheels are off the ground. Remove the front wheels so you'll have better access to the suspension.

Unbolting the shock at the bottom end is a little more difficult. It's bolted to the spring-seat rocker with two studs and nuts. To remove the nuts, come up through the bottom of the control arm with an extension and socket on a ratchet. Once the nuts are off, pull the shock out through the top of the spring tower.

Disconnect Tie Rods—As an integral part of each front-suspension spindle—steering knuckle—is the *steering arm,* the long arm that points to the rear of the car. Attached to the steering arms are the outer ends of the tie rods. The inner ends of the tie rods are attached to the *center steering link*—it operates between *Pitman arm* (arm on the steering gear) and *idler arm* (arm that's symmetrically opposed to the Pitman arm on opposite frame rail). Remove the nuts from the tie-rod ends. That's the easy part. Now for the hard part: Because of the tight fit resulting from the taper on the tie-rod end, it will be difficult to remove.

On page 59 is pictured a strange looking tool called a *pickle fork*. Get one and use it to separate the ball joints from the steering arms. You'll use it later to remove the spindle.

To use a pickle fork, wedge the pronged end between the tie rod and steering arm. Give the other end a sharp rap with a heavy hammer and the tie-rod end and spindle should separate. Unfortunately, this operation usually destroys the tie-rod-end dust

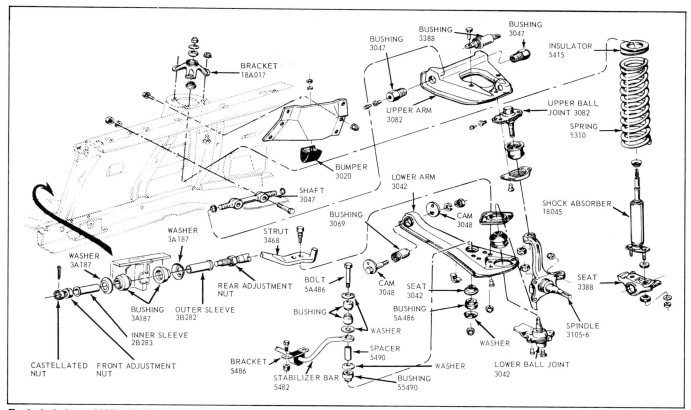

Exploded view of '67—70 Mustang front suspension; '65—66 suspension is similar. Drawing courtesy Ford Motor Company.

boot. If this occurs, the ball joint should be replaced. So consider doing the following.

If, for some reason, you can't get a pickle fork and you're going to replace the ball joints, use brute force to break loose the ball joint. Thread a nut on flush with the end of the ball-joint stud and beat on it with a hammer until the stud breaks loose. This may destroy the threads, so use this method only if you will be replacing the ball joint! You won't have any choice afterward.

When using a hammer, back up the steering arm by setting it on a block of wood or back it up with a brick-size block of steel. This will ensure that the full force of the hammer will drive out the ball joint rather than bend the steering arm.

Yet another, and better alternative is to push the tie-rod end out using a C-shaped puller. This tool is shaped like a "C" with a bolt threaded through it. The open ends of the "C" are slipped over the steering arm, and the bolt threaded up against the tie-rod-end stud. Tightening the bolt presses the stud out of the steering arm. Because the tool doesn't touch the tie-rod end except at the stud, the tie rod is not harmed. Also,

there is no chance of steering-arm damage with this tool.

Remove Springs—It's time to make friends with that guy at the spring shop who's going to re-arch your rear springs. Ask him if you could rent his coil-spring compressor. If he is unwilling to loan or rent it to you, you may be able to rent the tool from a major tool-rental company. No matter what, you need a spring compressor to remove the coil springs. The upper A-arm will not lower far enough to merely let the suspension down slowly with a jack. With the arm at full droop, the spring will still be captive.

Insert the bottom plate against the first bottom coil it will fit into. Adjust the plate so its groove fits firmly against the coil. Next, insert the top plate against a coil as close to the top of the spring as possible. Check again for full contact between the plate and spring.

Some spring compressors, including Ford's official tool, require that the top plate install on top of the spring tower. All this means is you must leave the spring compressed in place until the upper arm is

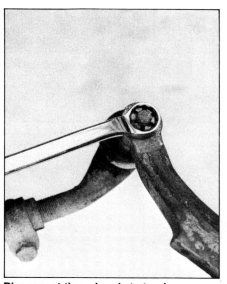

Disconnect tie-rod end at steering arm. After loosening nut, use pickle fork or lead mallet to break stud loose from arm. Support back side of arm if using a lead mallet.

With shock removed, spring compressor installs as shown. Do not attempt to remove upper arm with spring in place.

Ball-joint nuts are very tight—about 90 ft-lb. Use 1/2-in.-drive socket and long-handle ratchet or breaker bar to break loose nuts.

out of the way.

Insert the tensioning screw—it looks like a giant bolt—through the two plates and rotate the locking lugs into corresponding grooves in the lower plate. The upper locking lugs must also be secured. If it's a typical spring-compressor kit, enclosed is a part that resembles a long, 1/2-in.-drive socket.

With one end of the socket fitted to the tensioning screw, turn it with a 1/2-in.-drive ratchet. This will draw the two plates together, compressing the spring. When sufficiently compressed, remove the tool and spring from the car as a unit, assuming it's this type of compressor. **Be careful.** Handle the compress spring like you would a bomb . . . very carefully. Do not knock the tool, or let the assembly fall to the ground. A compressed spring has stored within itself a considerable amount of energy. To free the tool, reverse the ratchet and unload the spring.

Installed on top of the spring is an *insulator*. This is to prevent squeaks as the spring works in the spring tower during suspension movement. It may have come out with the spring. If not, reach up into the spring tower and pull it out. Don't lose this insulator as it should be reinstalled later.

Remove Drum Brakes—If your car has drum brakes at the front, partially remove each brake.

Start by pulling off each drum-and-hub assembly. To do this, pry off the dust cap from the center of the hub. Remove the cotter pin from the end of the spindle, then the castellated nut and washer. Wiggle the drum to dislodge the outer wheel bearing and remove it. Now pull off the drum. If the drum is stubborn, possibly because the drum is hanging up on the unworn inner edge or rust buildup on this edge, back off the brake shoes with the adjuster. This should provide the clearance necessary for the drum to come off.

Remove the shoes, springs and miscellaneous hardware, leaving only the backing plates and wheel cylinders. Turn to page 82 for details on how to do this.

Next, disconnect the brake hose from the brake line. Use a *flare-nut* wrench to break loose the lines. Otherwise, you may round off the fitting hex. To do this, pull off the clip that retains the hose end to the bracket. Then, disconnect the steel line from the hose end using flare-nut wrenches. If you're doing a ground-up restoration, it's best to remove the brake lines and their clamps later. For now, tape over the end of the line to keep out dirt and moisture.

Remove Disc Brakes—If your car has discs on the front, remove the calipers first.

There are two types of disc brakes used

on Mustangs: the four-piston, fixed-caliper style available through '68, and the single-piston, sliding-caliper used thereafter. The two are removed differently.

Four-piston calipers are super simple. Undo the brake line and two bolts that hold the brake to the steering knuckle and lift off the caliper.

Single-piston calipers are a little more complex. You have a choice of removing the caliper and its anchor plate as a unit—if you are going all-out on chassis detailing, or you can pull only the caliper and leave the anchor plate on the steering knuckle. The brake chapter has more on the pros and cons of this decision.

To pull the anchor plate and caliper, snip off the safety wire on the two anchor-plate bolts, remove the bolts and then the assembly. You'll need a stout breaker bar for the upper bolt as it is torqued to around 100 ft-lb.

If only the caliper is desired, remove the four bolts on the caliper inner side. Of course, with either method, the brake line must first be disconnected.

The final variation is to leave the brakes alone. If all you want to do is change suspension pieces, then remove the brake assembly as a unit. Just knock the upper and lower ball joints off with a pickle fork and set the steering-knuckle assembly to one side.

To remove the rotor-and-hub assembly, pry off the dust cap from the center of the hub. Remove the cotter pin from the end of the spindle, then the castellated nut and washer. Wiggle the rotor to dislodge the outer wheel bearing and remove it, Now pull off the rotor.

Caution: With the front-wheel drums or rotors off, the machined spindles are exposed. They are easily damaged from impacts from other tools or parts, or from moisture. To protect the spindles, tape a greasy wrag around each to cushion blows and seal against corrosion.

Remove Spindles & Control Arms—You can now remove each spindle. Start by taking off the upper ball-joint nut. Now, separate the ball joint from the spindle using a pickle fork or hammer. Separating this ball joint will be more difficult than the steering arm's because it is larger and access is more difficult.

Once the upper ball joint is broken loose, the upper-control-arm assembly will be free at the outer end. You can now remove it from the inner fender well. The two bolts that hold the upper arm and its pivot shaft in place can be reached from inside the engine compartment. When these bolts are removed on '65—66 models, it will free a

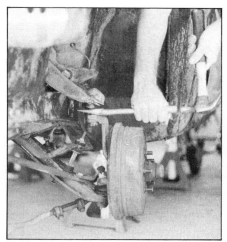

Pickle fork in action! Nut is loose, but in place to prevent lower assembly from falling and injuring your tender shins when stud breaks loose.

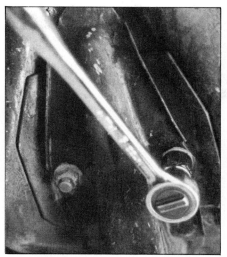

Nuts retaining upper control-arm-pivot shaft are on engine side of spring towers. Remove them to free upper control-arm assembly.

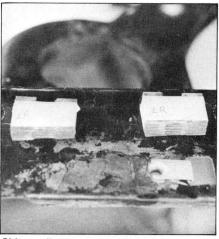

Shims adjust caster and camber on '65—66 models. They should be kept together in stacks and returned to their original positions so suspension settings will be reasonably close to spec. LF on tape is for left front; LR is for left rear.

group of shims that install between the pivot shaft and sheet metal inside the fender well. Keep the shims in their original groupings and label each so you can reinstall them in their original positions: FRONT or BACK and RIGHT or LEFT.

On '65—66 models, these shims are used to set caster and camber settings. When assembling the front suspension, reinstall them in their original positions and camber won't be too far off. (Caster is also determined by the adjustment nuts at the forward end of the lower control-arm strut.) This will allow you to get the car to an alignment shop under its own power without destroying the tires.

On '67-and-later models, caster is adjusted by lengthening or shortening the lower-control-arm strut rod via a pair of threaded nuts against the mounting bushing. Camber on these late Mustangs is adjusted by rotating an eccentric cam under the bolt head for the lower-control-arm inner pivot.

In the same fashion that you separated the upper ball joint from the spindle, separate the lower ball joint. Do this either with a pickle fork or by hammering on the end of the ball-joint stud. You can use some help here. Get a friend to support the spindle while you drive out the ball-joint stud. The instant it breaks loose, your friend will have his hands full of spindle. Have him set it aside.

Disconnect spindle from lower ball joint then remove.

Turn your attention to removing the lower control-arm assembly. Start by backing off the nut from the front of the strut—the hocky-stick-looking affair that runs from under the front crossmember to the outer end of the control arm. Don't move the nut at the rear of the bushing. It should be in the

Strut rod and stabilizer bar are attached to lower arm. Start by disconneting stabilizer-bar linkage. Some penetrating oil on threads will ease removal of these fasteners.

same place on the strut at assembly time so caster won't be too far off.

If a cotter pin is used to secure the front nut, remove it first. At the other end of the strut, remove the two bolts holding it to the lower arm. Slide the strut rearward out of its bushings to remove it. Remove the

Easiest way to remove upper-arm-shaft bushings is with impact wrench. If you don't have one, use a long breaker bar.

Clean, sand or beadblast and paint or powder coat upper arm and backing plate. Clearcoat spindle; it was unpainted from the factory. Don't paint it black as shown. Use a quality catalyzed enamel for job.

bushings, sleeves and washers, and put them back on the strut for storage.

Next, disconnect the stabilizer-bar link and the assorted rubber bushings and washers. Although the links should be replaced, reassemble them loosely so the parts won't get lost and they'll be kept in the correct order. You can now detach the lower arm from its inner pivot. On '67-and-later models, remove the camber-adjustment bolt/cam assembly and free the control arm.

Remove Stabilizer Bar—Once you have both control arms out of the way, remove the stabilizer bar. Remove the bracket from one side, lower the bar at that side, then remove the bracket from the other side to free the bar.

While you're underneath, remove the tubular crossmember that's between the lower-control-arm brackets. There's a bolt at each end of this crossmember. Remove them and lower the crossmember.

Remove Bump Stops—Finally, remove the front-suspension bump stops. There's one at each side, mounted under the outboard side of each spring pocket, which can also be removed. Do so if you're doing a ground-up restoration. Otherwise, just remove the bump stops. The mounting nut is probably buried under an accumulation of undercoating, dirt and grime, which you must dig out with a screwdriver or whatev-er to expose the nut. Once you've done this, use a six-point socket and ratchet to remove the nut and bump stop.

You have now removed the front suspension, assuming you've done both sides. You can jump ahead to Chapter 5 if you want and remove the steering, either power or manual.

This will complete the disassembly of the entire front of the car. Assuming that the front of the car is down to bare bones, you can paint the engine compartment and inner wheel wells.

DISASSEMBLE & RECONDITION FRONT SUSPENSION

At each end of the upper-control-arm shaft is a bushing that threads into the control arm. To remove these you'll need an impact wrench or 18—24-in. breaker bar. Remove one bushing, then thread out the shaft. Now remove the other bushing from the control arm.

A problem you may run into is destroyed A-arm bushing threads. The large bushings are made of thick steel and are quite strong. The thin A-arm material is no match for the bushings in a thread tug-of-war, and the A-arm often loses its threads during bushing removal.

If this happens, there won't be enough material to tap, even if you could find a tap that large. That leaves but two ways out: A-arm replacement, or welding.

Inspect the A-arm closely for cracks. Look around the rivet holes where the ball-joints mount. Also look for cracks radiating from the bushing holes, or running acrosss the A-arm anywhere along its edge. If you see any cracks, replace the A-arm. This does away with your bushing thread problem because the new A-arm should have good threads.

If the A-arm passes inspection, then consider welding the bushings in place. Do this during assembly, after you've centered the shaft in the A-arm. Tack-weld the bushings to the A-arm in several spots. Despite the bushings' impressive size, they don't carry loads a few tack welds can't hold. Also, if the bushings ever need to come out again, you can cut through a few tack welds, where a full-length weld would make things permanent.

Remaining on each control arm are the ball joint and spring seat. Two bolts and nuts retain the spring-seat rocker shaft to the control arm. Remove these and lift off the spring-seat assembly. To remove the ball joint, simply unbolt it from the control arm . . . if bolts are used. If the ball joint is riveted to the arm, cut off the rivet heads using a large chisel and hammer. Take care not to damage the control arm. The upper arm should now be bare.

The lower ball joint is riveted to the lower arm. Rather than replacing just the ball joint, discard the entire arm and ball joint and replace it with a new ball-joint-and-control-arm assembly. That's the way the ball joint comes both from Ford and most aftermarket suppliers . . . with the control arm. You can purchase ball joints and inner-pivot bushings separately, but when you add in the price or trouble of removing the old parts and replacing them with new ones, why bother? Get the complete new assembly.

You now have two upper arms, two backing plates—if drum brakes are used—two spindles and two coil springs to be cleaned and painted or powder coated. As for the remainder of the parts, discard them and purchase new ones to replace them. So, having done this, turn your attention to assembling the reconditioned and new parts.

Urethane Bushings & Other Tricks—A growing trend in Mustangs is restoring a car to look stock, or very close to it, while incorporating performance improvements. Such combinations of restoration and modification are called *restification* and such cars have been *restified*—not restored.

Another variation of this theme is to restore the car using aftermarket accessories commonly installed when the car was new. A Holley carburetor, set of headers, Hurst shifter and American Racing Equipment wheels give the car improved performance, yet still recalls the muscle-car era when these cars were new.

Obviously, if you are restoring your Mustang for show, both approaches are incorrect, but if you are building a Mustang to drive, or enjoy increased performance—why not? A few small changes make all the difference on how your car handles and performs—and most modifications have been part of the Mustang scene from the beginning.

The modification that concerns us now is urethane bushings. The rubber in the stabilizer-bar mounts and linkage, along with the rubber strut-bar mounts allow the suspension to *deflect* in response to cornering and bumps. Urethane bushings deflect less because they are much harder than rubber. The car responds better, and provides the driver with more feedback, or road feel. Yet the urethane bushings do have some compliance for really big bumps.

About the only two bad things you can say about urethane bushings is cost and squeaks. Urethane is more durable than rubber and is made in smaller quantities. Therefore, it costs more. On the other hand, the bushings shouldn't wear out—ever. Urethane bushings are also available in black, just like rubber bushings. You no longer have to use red or yellow bushings and advertise to the whole world you've

Assemble upper arm with new ball joint, spring seat and pivot assembly.

made your car non-stock. Horrors!

What they might do is squeak. The harder material sometimes squeaks when the stabilizer bar moves against it during suspension movement. An occasional shot of silicone lube will stop almost any squeak. But if you are reshooting the movie *Bullit* in your fastback, you might hear a squeak as the suspension goes through full jounce and rebound. Also, lifting the car on a hoist can cause a whole lot of "haunted-house" creaking.

The choice is yours. Installation is the same for original rubber or modern urethane. But now is the time to go one way or the other; while the suspension is all the way off.

Another thought is installing a larger front stabilizer bar, a rear stabilizer bar, traction bars in the rear or maybe even high-rate springs. Again, cost is a factor, but the handling improvements are impressive. Also, some of these modifications require a hole or two to be drilled. If drilling holes in your car sounds like too much modification—even if no one will ever see them under the car—then don't consider these mods.

Relocated Upper A-Arm Pivots—While we are mentioning suspension mods and drilled holes, there is a common mod you should know about: drilling a new pair of

holes in the inner fender for the upper A-arm mounting bolts, exactly 1-in. below the stock ones. First, if you like a lowered car, it lowers the front about 3/4 in. Second, and more importantly, this makes the upper A-arm swing farther inboard when the suspension travels into jounce—as the outboard side does when going around a corner. This increases the rate negative camber is added into the suspension. The result is increased tire "footprint" and, consequently, cornering power. Carroll Shelby made this modification to the '65 and '66 G.T.350s. Only when Ford told him to at least break even in '66 did he give up this extra step, midway into the model year.

The big disadvantage is you have to drill four holes in your car, two per side. If you want a car stock, this mod isn't for you. And while anything is possible, filling in these holes at some later date to return to stock is a royal pain. Again, the decision is yours.

ASSEMBLE & INSTALL FRONT SUSPENSION

Install Upper Control Arms—Get two new upper arm-shaft bushings. Be sure each has a 90° grease fitting in the end. If the fitting is straight, you won't be able to get a grease gun on it. Plugs are frequently used instead. If this is the case, replace the plugs with grease fittings. But don't fully tighten them now.

Included with the upper-control-arm shaft and bushings are two O-rings. These are inserted, one into each bushing. You'll see the groove for each O-ring just inside the bushing. Install the O-ring so grease won't run out between the shaft and bushing. Now, thread one of the bushings into the control arm and torque it 35—45 ft-lb.

Thread the shaft into the bushing until it will go no further, then insert the remaining bushing and torque as above. Rotate the shaft into the newly installed bushing until the shaft holes are spaced the same distance from each side of the arm. This centers the shaft in the arm.

Install the upper ball joint next. Position the ball joint in place with its dust boot, secure it loosely with the bolts and nuts supplied with the ball joint, and torque the nuts 30 ft-lb. (When I give an exact torque, it is always plus or minus 5 ft-lb.) Finally, install a new spring-seat assembly using the serrated bolts from the original spring seat. Note: The spring-stop tab goes toward the rear of the arm. There is a 90° metal tab on the side of the spring seat (part 3388 that supports the spring and shock absorber in the lower arm [part 3042]) that prevents the

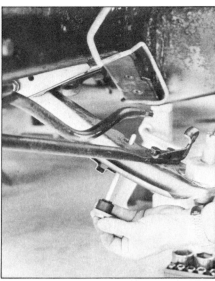

Upper arm completed and installed: Note 90° grease fittings in ends of upper-arm bushings.

Once stabilizer bar is bolted to underside of front members, loosely install links to each lower arm. Once both are installed, tighten links. Note new lower control arm with riveted-on ball joint.

spring from rotating in its seat.

The rear of the arm faces away from you when installed—or the pivot point of the arm. Torque the nuts 30—40 ft-lb. Once you've done both, reconditioning of the upper control arms is complete and they are ready for installation.

Loosely install the upper control arm, then install the shims between the pivot shaft and sheet metal over the same bolts on which they were originally installed. If you didn't label them during disassembly, you'll now wish you had. Torque the nuts 75—100 ft-lb.

Now's a good time to install new bump stops. The olds ones are probably cracked and deformed. Fit the stud up through the hole in the bottom edge of the bracket, place a lock washer over the stud and thread on its nut. Torque the nut 12—20 ft-lb.

Install Lower Control Arms—Most, but not all, Mustang lower control arms are sold with the lower ball joint as a unit. If you purchased one of these, simply bolt the lower arm to its bracket. Line up the bushing sleeve with the hole in the bracket, then slide in the bolt with cam in place from the front. On the other side, fit the other cam and lock washer to the bolt and thread on the nut. Tighten it 75—100 ft-lb. If you purchased replacement lower ball joints separately, first attach the ball joint to the lower arm with its three nuts and bolts, using locking compound on the threads. Torque the nuts to the specs supplied with

the ball joint. Then install the lower arm.

Turn your attention to the lower-control-arm strut. Using new bushings, install in order on the threaded end of the strut the inner adjustment nut, outer sleeve, washer and large rubber bushing. Thread the nut down to its original position from the end of the strut. From behind the bracket hanging below the radiator support, pass the rod through the bracket at the threaded end, and lay the flat end on top of the lower A-arm.

Now install the two bolts through the flat end of the strut and into the lower control arm. Torque it 5—11 ft-lb.

Back at the front of the strut, install the remaining sleeve, rubber bushing, large metal washer and outer adjusting nut. Torque the outer adjusting nut 60—80 ft-lb.

Because you've changed the suspension settings completely, don't expect this final torque to result in an accurate caster adjustment. When your car is driveable, take it to an alignment shop where the front end can be set up perfectly. In other words, don't worry about getting it spot on the first time.

Install Stabilizer Bar—The last item that bolts to the lower control arms is the stabilizer bar. If you haven't checked the stabilizer-bar bushings, do it now. Replace them if they are mushy, split or cracked.

Note where the bushings are and remove them from the bar. Slit the bushings along the length of the bar with a knife and pull them off. That's easy. The hard part is

installing the new ones. You'll need some lubrication—a bucket of soapy water or a can of spray silicone. *Don't use oil!* Lubricate the bar, then force each bushing over the end of the bar. Check their positions on the bar by holding the bar up under the bracket mounting points. Or, measure the distance between the bracket bolt holes and space the bushings the same. Either way, adjust the bushing accordingly and position them the same distance from the center of the bar.

You can now install the bar. Place the outer ends on top of the control arms, raise the bar to the underside of the boxed members, and secure it with the two brackets. Torque the bolts 17—30 ft-lb.

Finish installing the stabilizer-bar assembly by fitting the new links and bushings between the bar and lower arm. These are available in kits in almost any parts store. A bushing and washer installs on both sides of the bar and arm, the spacer goes between these two sets of bushings and washers, and the bolt installs from the top down. Install the nut and tighten it so all four bushings are correctly seated. Torque the link nut and bolt 5—12 ft-lb. Connect the other end of the bar to its arm and you'll have both pairs of control arms installed. You can now turn your attention to the backing plate and spindle.

Install Drum Brakes—Begin assembling the drum brakes by bolting the backing plates to the spindle. Torque the mounting bolts to about 30 ft-lb. Now, bolt a front wheel cylinder to each backing plate. If both front and rear wheel cylinders are off and you're not sure which goes where, the front cylinders have larger pistons. Position the wheel cylinder so the boss for the line fitting goes to the rear of the car.

Once you have the cylinder fitted to the backing plate, torque the bolts 10—20 ft-lb (10-in. drums) or 5—7 ft-lb (9-in. drums). You can now mount spindle to the upper and lower arms.

Turn to page 85 to finish assembling the drum brakes.

Install Disc Brakes—Don't bother with the disc brakes now. Instead, wait until after you've installed the front suspension, then turn to page 92 for details.

Note: If you have a spring compressor that pulls the spring up against the bottom of the spring tower to compress it, install the spring now rather than after the spindle is in place. Just make sure the spring insulator is between the squared end of the spring and its seat in the spring tower.

Install Spindle—Taking care not to damage the machined portion of the spindle, install it between the upper and lower ball joints. Torque the nuts 60—90 ft-lb, then back off each until the castellations align with the cotter-pin holes. Install new cotter pins and lock them in place. Bend the long cotter-pin leg over the threaded end of the stud. Bend the short leg down over the nut and clip it off.

With the steering linkage in place, page 68, you can slip the tie-rod-end stud into the steering arm and secure it with its nut. Torque the nut 35 ft-lb and continue tightening it until the first castellation aligns with the cotter-pin hole. Install the cotter pin and bend it over as you did with the upper and lower ball joints.

Install Springs—Once both spindles are in place, install the coil springs. You may need some help from a friend to do this.

Using the spring compressor, compress the spring. This is where you may need the help. It takes two men and a baby donkey to hold the spring while you compress it. When doing this, remember that the *squared end*—it's flat—installs up into the spring tower. Once it's sufficiently compressed, place the insulator on top of the spring and position the assembly against its seat in the spring tower. Rotate the spring so its free end at the other end butts against the tab on its seat on the lower control arm. Gradually back off on the spring compressor, making sure the spring seats properly. If all looks OK, turn your attention to

the shock absorbers.

Install Shock Absorbers—Now's time to break out a set of new front shock absorbers. Lay out the shocks and their bushings, washers and nuts.

Place a bushing over each stud on the shock-absorber mounting plate at the bottom of the shock. The stepped end of each bushing goes away from the plate. Pull the shock-absorber shaft out to its fully extended position. Insert the shock bottom end first through the top of the spring tower and onto the spring seat. Rotate the shock so the mounting studs align with the holes in the spring seat. You may have to lever the spring seat up at its outboard end so the studs will drop through the holes. Hold the spring seat in position while you install bushings, washers and nuts on the studs.

To finish installing the shocks, wait until you can set the front end on the ground. This will compress the front suspension so the upper end of the shock will extend farther out of the top of the spring tower, allowing you to easily install the upper mounting bracket.

Completely restored and assembled front suspension, brake and steering less shock absorber, bumpstop bracket and dust cap.

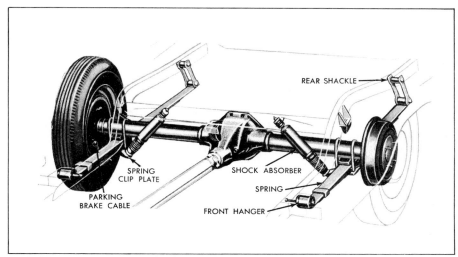

Basic Mustang rear-suspension layout: After you've finished with the front suspension, move to the rear. Drawing courtesy Ford Motor Company.

Take the opportunity now to lubricate all of the ball joints and upper control-arm pivots. Don't force more grease in than necessary. The ball-joint dust boots should only be firm. Anymore and you risk bursting a boot. The only way to correct such damage is to replace the ball joint.

Install Brakes—Regardless of whether your Mustang is equipped with drum or disc brakes, install them now. This means installing the wheel hubs with freshly greased new or cleaned wheel bearings. For details on installing brakes, turn to pages 85 or 92.

63

With wheels off the ground, axle housing must be raised to unload springs.

Parking-brake cable housing is removed from backing plate by compressing spring fingers (arrow).

Accessory stabilizer bar is bolted to underside of axle housing. Remove nuts from U-bolts, and lower bar. Free ends attach to body via links similar to those on front bar. Disconnect them to remove bar.

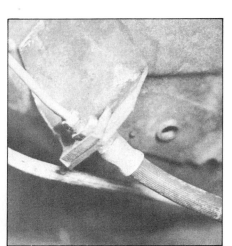

Disconnect rear-brake flex line as you did the front. Coat with penetrating oil, undo flare nut with flare-nut wrench while holding large hex with another wrench. To remove hose from bracket, grasp U-clip flange with pliers and pull out clip.

To complete the front-suspension installation, connect the hose to the wheel cylinder or caliper. Remember to use a flare-nut wrench to avoid rounding off the shoulders of the fitting. Attach the other end of the brake hose to the steel line and clip the union to the mounting plate.

REMOVE REAR SUSPENSION
Rear Axle—Using your floor jack, raise the rear axle *just enough* to take the load off the tires—no more. Support the car with

jack stands under the rear torque boxes. Check that the car is stable.

Once you have the wheels off, remove most drum-brake components. Start by removing the *Tinnerman*—sheet-metal—nuts from the wheel studs. Loosen the nuts with a pair of pliers, then finish removing them with your fingers. Pull off each drum. This will expose the drum-brake innards which you can now remove.

Remove the brake shoes and associated parts from the backing plate. Once you've

taken these brake parts off, turn your attention to the parking-brake cable. Disconnect the cable housing where it snaps into the backing plate. This must be done before you can remove the axle.

Notice the spring tabs on the end of the cable housing. Compress these tabs together and pull the cable from the backing plate. This is sometimes easier said than done. While pushing on the end of the cable housing, push each tab toward the hole in the backing plate with a screwdriver. Work around the end of the housing until the end of each tab is in the hole and you can push the housing through. Disconnect the other cable and wire them up out of the way. Now, hop on your creeper and roll under the car.

On cars with rear stabilizer bars, remove the bar. You can either undo the bar ends directly at the body, or where the body-to-bar link joins the bar proper. Then on the axle housing, remove the clamp bolts and the bar is off.

On all cars, unbolt the drive shaft from the rear axle. Before making any disconnections, mark the drive-shaft-to-axle relationship with chalk, paint or punch marks. Punch marks are best because they are permanent. Now remove the nuts and U-clamps from the rear U-joint and let the shaft gently down onto the floor. Do not drop the drive shaft!

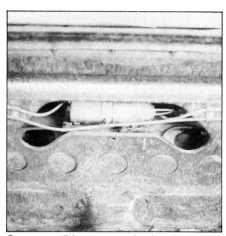

On convertibles, upper shock mounts are accessed through hole in reinforcement behind rear seat back. Convertible-top pump is visible through this hole. Back off each mounting nut while a helper underneath keeps the shock from turning.

Remove axle-housing breather-tube connector to free rear brake-line junction block and lines from axle housing.

To remove axles from housing, remove retaining bolts and nuts. This will also free each backing plate. You may have to jerk its flange or whack it with a large mallet to break loose the axle.

If you want to remove the drive shaft completely, creep forward and support its transmission end. Gently slide the shaft backward until the front slip yoke is free of the transmission. Transmission fluid will probably leak out of the rear seal. A rag, or special rear seal plug made from an old slip yoke should stop the leaking. It's also a good idea to change the seal if it shows signs of leaking while in service. Store the drive shaft in a safe place where it can't get dropped or bent. If it does, it will vibrate and need to be replaced.

Disconnect the shock absorbers at the spring plate. Push them aside for now. Remove the nuts from all four U-bolts, then the U-bolts and spring plates. This will free the axle from the springs. If the nuts and U-bolts are so badly rusted that you can't get them off, you may have to "torch-off" the U-bolts.

Now that the axle is loose from the springs, scoot forward and under the drive-shaft tunnel. Just above you is the brake hose that runs from the brake line in the tunnel to the junction block on the axle. Disconnect this hose at the mounting bracket just as you did on the front brakes. Now, go back to the very rearmost part of the car.

Double-check that the jack is still supporting the axle, but no more. If your floor jack bleeds down the slightest, you may have more than you can handle during the next few steps.

Remove the nuts from each rear shackle assembly and slide the shackle from its bushings. Lower the end of the spring. Watch that the axle doesn't fall off the jack. Drop the other spring and roll the axle out from under the car. Do this with the jack or mount the wheels (with tires) to the axle.

Remove Springs & Shocks—All that remains to be removed at the rear suspension are the leaf springs and shocks.

Remove the mounting bolt from each front spring eye. Remove the nut and slide out the bolt. Set the spring aside and remove the other spring.

The shocks are more difficult to remove. You'd think the Ford engineers tried to hide the top shock mounts from you. If, however, you remove the rear seat back from the convertible, rear floor from the fastback, or floor mat from the trunk of the coupe, you'll find these mounts. On the convertible, they're on either side of the convertible-top pump.

Although you can do it yourself, it's best to enlist help from a friend when removing the shock absorbers. As with the bottom mount, a bayonet-type mount is used at the top of each shock. Consequently, if you turn the nut, the stud will turn in its bushings.

While you're up top removing each nut,

Notice different lengths of axles. Long axle goes to passenger's side.

have your friend underneath hold the shock to keep it from turning. If he has trouble with it turning, have him grasp the shock body with a pair of Channellock pliers. Once the nut is off, the shock can be pulled down out of its mount.

Strip Axle—The easiest way to work on a rear axle is to set it on two jack stands, one at each end. Do this and remove the wheels if you reinstalled them.

Remove the steel brake line that runs between the junction block and both rear wheel cylinders. Unbolt the junction block, pry up the tabs that bend over the brake line, and lift the assembly from the axle housing.

Next, remove the four nuts that retain the axles to the housing. Rotate the axle so the large hole in the axle-flange hole aligns with a nut. Remove that nut with a socket wrench and rotate the axle to the next nut. Once the nuts are off, pull the axle shaft from the housing. Then remove the backing plates. Remove the other axle shaft and backing plate. Note that the passenger's side axle is longer than the other.

Unless you suspect that the ring and pinion gears, differential gears, their bearings or Traction-Lok (if so equipped) need attention, don't bother removing the center section (carrier assembly), if it's removable, from the axle housing. However, if

you suspect there is a problem, as would have been indicated by leaks or excessive noise, take the axle assembly to a professional mechanic. He'll have the tools and knowledge to do the job.

RECONDITIONING & REASSEMBLY

A complete restoration of your car should include re-arching the rear springs as discussed earlier in this chapter. This will ensure the car will have the correct ride height at the rear. So, load up the springs and deliver them to the spring shop.

If you want the springs to look their best, after the leaves have been re-arched, pick them up from the spring shop and paint them yourself or have them powder-coated. Then, return them to the spring shop for assembly. The paint job spring companies usually do is not very good.

Now, consider the axle bearings. If the

rear axle was noisy or has high mileage on it, replace the bearings. Take the axle shafts to an auto-repair shop and have the old ones pressed off and new bearings pressed on. The seals in the axle housing get replaced, too.

As for things you can do yourself to the rear axle, they're next. Start by painting the axle housing semi-gloss black. First, remove all the grease, dirt and grime, wipe it down with lacquer thinner, then sand the housing. Paint or powder-coat the brake backing plates black. Now is the time also, to paint or undercoat the bottom of the floor pan and rear wheel wells. If you elect to paint, after doing any body repairs and prep work of course, use the same color as you will use for the body exterior.

INSTALL REAR SUSPENSION

Assemble Rear Axle—Begin by mounting the wheel cylinders to the backing

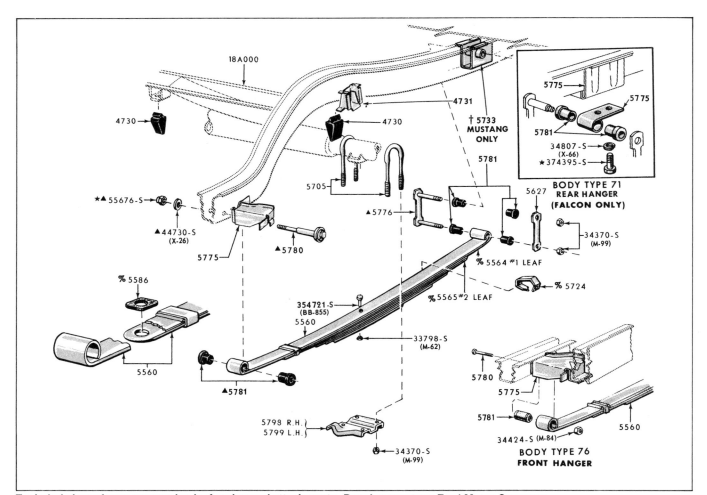

Exploded view of rear-suspension leaf spring and attachments. Drawing courtesy Ford Motor Company.

plates. The brake-line nipple points to the rear. Fit the backing plate to the axle and insert the axle shaft into the housing. Be sure to use a new gasket between the mounting flange and housing.

Rotate each axle into engagement with its differential gear. When the splines engage, the axle will seat. Now, install the four bolts and torque the nuts 25—35 ft-lb. Remember that the short axle goes to the driver's side; long axle to the passenger's side.

Finish assembling the axle housing by installing the brake lines. Either clean up the old junction block-and-brake line assembly or use it as a model to make a new set of lines. If you make new lines, use *steel tubing with double-flared ends.*

Install Springs—Attach the leaf springs to the front hanger. The larger eye goes to the front. If you have any doubts, the spring clamping bolt is closer to the front eye than it is to the rear eye.

Raise the front spring eye to the front hanger and secure it with the bolt and nut. Torque them 30—50 ft-lb on '65—68 models and 70—90 ft-lb on '69—70 cars. With the springs hanging down on the floor, roll the axle underneath. Do this with the axle balanced on a floor jack or with the wheels mounted on the axle. Raise the axle with the floor jack until the rear of the springs can be raised high enough so they can be secured at the rear with the shackles.

With each shackle on the body and fitted with new bushings, raise the spring and slip the spring eye over the shackle bolt. With all bushings in place, fit the shackle plate over the shackle bolts and thread on the two nuts. Torque the nuts to about 22 ft-lb. You can now lower the axle onto the springs.

Install Axle—Set the axle on the springs so its pads pilot over the spring clamping bolts. This positively locates the axle to the springs. You'll probably have to move the axle around a little so it drops into engagement over the clamping bolts. The spring pad should be flat against the spring.

Hang the U-bolts over the axle, one on each side of each spring. With nuts within easy reach, fit the spring plates to the U-bolts and thread on the nuts. Run the nuts down gradually and torque them 30—50 ft-lb.

Install Shocks & Brake Lines—Unless the shocks were brand new when you removed them, install new ones. Extend the shock-absorber shaft, slip a washer and bushing over its end, and insert the end of the shaft up through the mounting hole in the floor. Up top, have a friend slip the other bushing half over the shaft, followed

If your Mustang's differential clunked or whined in service, better take it to a differential specialist. Otherwise, clean then paint housing with same catalyzed enamel as used on front suspension.

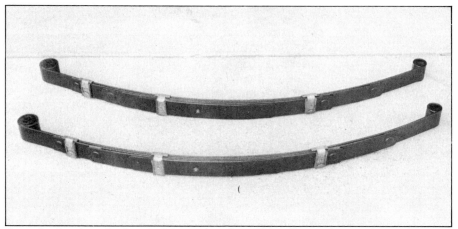

Re-arched springs fresh out of the spring shop: This will cure the sagging-spring problem.

by the washer and nut. Torque the nut to 15—25 ft-lb. You'll have difficulty in using a torque wrench for this, so guesstimate. At the bottom end, use the same process to attach that end to the spring plate.

Using flare-nut wrenches, connect the brake hose to the steel line. Clip the brake-

hose union to the mounting bracket. Now, you can get out from under the car and connect the parking-brake-line housing to the backing plate. Push the cable sleeve into the back of the backing plate so the locking tabs snap into place. Be sure the two tabs are expanded and resting against the shoulder. Pull on the sleeve to check.

5

Steering

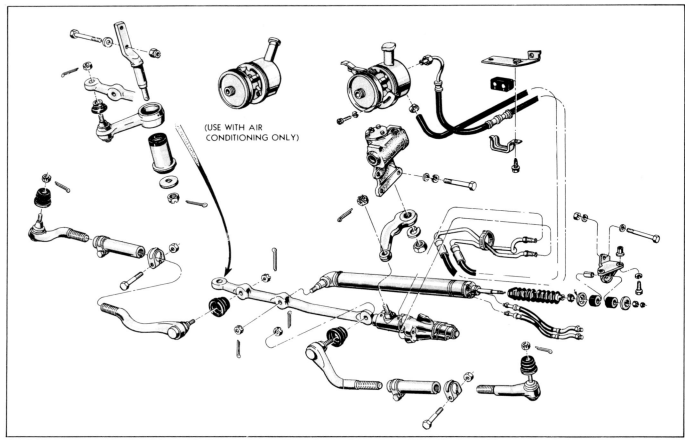

Assist-type power steering (shown) is the most complicated chassis system of the Mustang. The reverse is true of the manual system; it is very simple. Courtesy Ford Motor Company.

Mustangs covered in this book use a recirculating-ball-type steering gear. We'll look at both manual and power-assisted steering in this chapter. However, because power steering is the most complex, but not that different except for some additional teardown and assembly steps, I'll use it as the example.

Mustang power steering is the *assisted* type, so there are only two major differences between it and the linkage used with manual steering: the addition of a *power cylinder* between the bottom of the left

"frame rail" and the center steering link; and the *control valve*—it tells the power valve how much to assist the steering—mounted to the end and between the center link and Pitman arm.

Steering gears for both setups are identical in design, but the gear used with power steering is *quicker*—there are fewer turns lock-to-lock. Therefore, servicing these gears is identical. Of course, the remaining difference between the two types of steering systems is the addition of a hydraulic pump that supplies pressurized

fluid to the power-steering control valve and power cylinder.

In the chapter on suspension just covered, we separated the suspension from the steering at the tie-rod ends. So, that's where we'll pick up.

Before you removed the suspension components, though, you should have double-checked the steering and noted any problems. For instance, excess play in the steering indicates a need for adjusting steering-gear backlash or replacing worn-out tie-rod ends. Also, if the front-end

Bird's eye view of power-steering components from back of engine compartment.

Replace idler arm. You can't rebuild it.

hunted—wandered—and the right front tire wore more than the left, chances are the idler arm is worn out.

Adjust Steering Gear—There is but one adjustment procedure to keep the recirculating-ball-nut assembly and sector gear properly adjusted. The adjustment keeps end play and backlash to a minimum.

Many people take a look at the locknut and adjusting screw in the steering-box cover and figure all they need to do is tighten the adjusting screw to reduce backlash—or play—in the steering. Doing so may reduce play somewhat, but it will also cause premature wear of steering-box internals due to too-little clearance.

The correct procedure takes a little time and effort, but is the only way to get the steering into shape. If you don't have the tools, have an alignment shop do this job.

Start under the car at the Pitman arm. This is the curved arm connecting the shaft dropping out the bottom of the steering box to the steering linkage. Disconnect the Pitman arm from the steering gear with the tool shown on page 70. This eliminates all drag in the steering linkage from your measurements.

Now go up to the steering box and locate the sector-shaft adjusting screw and locknut. Loosen the locknut and back off the adjusting screw a few turns counterclockwise. This is easier said than done on some big-block models, but a long screwdriver

and bent wrench make it possible.

Inside the car, strip the steering-wheel horn ring and trim off until the large nut holding the steering wheel to the steering shaft is visible. Connect a socket and in-lb torque wrench to this nut. You need a beam-type torque wrench to read how much torque you are applying to the steering-shaft nut.

Starting with the steering a little off-center, turn the steering shaft via the torque wrench 1-1/2 turns. Read the amount of torque necessary to turn the steering. It should be 4—5 in-lb. If there's a SMB-K model number on your car's steering-box tag, try for 3—4 in-lb. However, with this quick-ratio (16:1) steering, the 4—5 in-lb figure is specified by Ford for "improved or competition handling."

This reading is the worm-bearing pre-load. If it is to specification, the steering is properly adjusted. There is no need for further adjustments, so you can continue with cleaning or parts replacement elsewhere in the system as necessary.

If worm-bearing preload is not to spec, there are two reasons why it might not be. First, the system is out of adjustment, and needs to be reset. Second, worn parts could be causing excessive clearance. After 20 years, count on the latter being the case. Readjustment may or may not bring the system into spec, depending on how worn the parts are. But no matter what the rea-

son, readjustment is the next step in determining what is at fault.

Go back to the steering box and locate the bearing-adjuster locknut. This is the very large nut on the end of the steering box where the steering column enters. The locknut is about the diameter of the steering box. Loosen it with large, slip-joint pliers, or a hammer and punch held in the nut's depressions. Now loosen or tighten the adjusting nut that fits right around the steering column.

Go back to the steering wheel and re-measure the turning load. Keep tweaking the adjusting nut until you get the desired 4—5 in-lb (or 3—4 in-lb with SMB-K steering). When you've got the adjustment in range, tighten the locknut and final-check the preload with the torque wrench.

Now turn the steering wheel to either stop. Do this slowly so as not to harm the ball return guides in the steering box. From the stop, reverse the wheel 2-3/4 turns to center the ball nut inside the steering box.

Tighten the sector adjusting screw—the one on top of the steering box—until it takes 8—9 in-lb to rotate the steering wheel past the ball-nut center spot. The 8—9 in-lb figure represents the preload on both the worm bearing and sector shaft—called the *total center meshload*.

Once total center meshload is set, tighten the sector-shaft locknut and double-check the load. The steering is now adjusted. You can reconnect the Pitman arm.

If all went well, there is not enough wear in the steering box to worry about. But if any of the preload checks did not come to specification, the steering box should be replaced or rebuilt. See the section on rebuilding steering boxes at the end of this chapter, page 78.

Idler Arm—Normally, idler arms are replaced along with all the other moving parts of the steering system during a restoration. But if you've replaced the idler arm in the recent past, blindly replacing the idler arm may be more a waste of money than sound preventive maintenance.

Idler-arm wear is easily detected by grasping the end of the arm and levering it up and down. If movement is noticeable, the idler arm is definitely shot and needs to be replaced.

Use this style of ball-joint separator rather than pickle fork to remove ball-joint studs from idler arm and Pitman arm. Using pickle fork on power-steering unit could damage control valve.

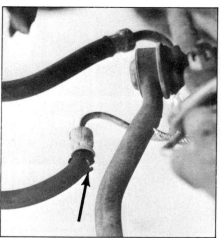

Save power-steering hoses only for making sure you get the correct replacements. Damage (arrow) is typical.

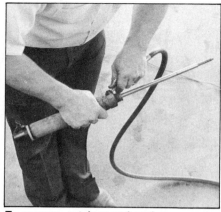

To remove, retainer and seal may need such encouragement. Compressed air will drive out shaft, carrying these two components with it. Point shaft where it will do the least amount of damage when it comes out.

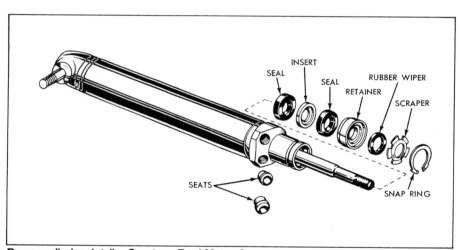

Power-cylinder details. Courtesy Ford Motor Company.

STEERING-LINKAGE DISASSEMBLY

Loosen the clamping sleeves at both inner tie-rod ends. Then remove each tie rod by unthreading it from the inner tie-rod end. Leave the inner tie-rod ends hanging from the center link. If the car has power steering, the center steering link is part of, or attached to, the power-steering control valve.

Don't use a pickle fork to remove the center steering link from the idler arm or Pitman arm. You'll damage the center link if you do. Use the tool shown at the top of this page to do this.

If your car has power steering, it will have a *control valve* mounted to the left end of the center steering link and the Pitman arm. There will also be a *power cylinder* between a bracket that's under the left "frame rail" and the center of the steering link. Two hoses connect these units.

First remove the power cylinder from the center link. Disconnect the hydraulic line at the control valve and let the fluid drain into a pan. At the end of the power-cylinder rod, remove the bracket from the underside of the "frame rail." At the center-link end, remove the cotter pin and nut. Use the puller to remove the cylinder from the center link.

At the control valve, disconnect the two lines coming from the power-steering pump. Let the fluid drain into a pan. To flush fluid from the valve, turn the steering wheel from side to side several times. At the center-link end of the valve, loosen the clamp and remove the roll pin from the center link through the slot in the control-valve sleeve. At the Pitman arm, remove the stud nut, then then force out the stud using the tool pictured. Now, turn the wheels fully to the left and unthread the control valve from the center link.

To complete the steering-linkage disassembly, remove the idler arm with the center link from the right "frame rail." At the other side, remove the steering gear. Like other major components, steering-gear work requires special equipment and experience. For instance, a two-ton hydraulic press is needed to remove and install new bushings. So farm it out when major work is required.

If your car has manual steering, skip the following discussion. Clean up, restore and reassemble the steering linkage. Also, remove the steering gear, manual or power, and rebuild it or, if it's in good condition, give it a meticulous cleaning. If there's too much slop, adjust the sector-shaft preload as described earlier.

Replace the tie-rod ends and idler arm. If the Pitman-arm ball joint is worn out, replace the Pitman arm. Use a puller to remove it from the steering gear. If you are

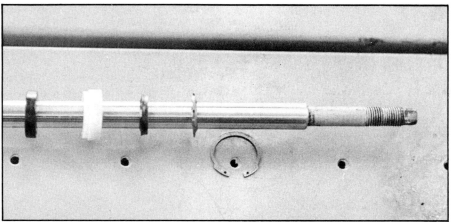

From left to right, parts you'll get in power-cylinder rebuild kit include: seal, retainer, wiper, scraper and retaining ring. They install in order shown.

Be sure spring (arrow) inside seal installs *into* cylinder.

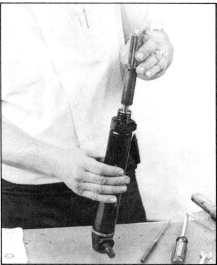

Jerry-rigged deep-well socket and a hollow gear is used to force scraper into its seat.

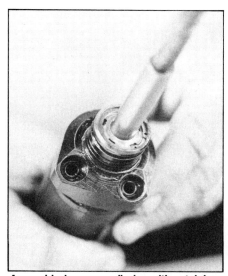

Assembled power cylinder with retaining ring in place.

going to rebuild the power-steering components, read on.

RECONDITION POWER CYLINDER

Disassemble Power Cylinder—Begin the power-cylinder disassembly by removing the bracket from the end of the shaft. Slide off the dust boot. Save the bracket, but discard the bushings and dust boot. These will be replaced.

Note: When handling the power cylinder, be careful. If its bore is scratched or dinged, even the best seal will leak.

Using a pair of retaining-ring pliers, remove the retaining ring—commonly called *snap ring*—from the power cylinder. Next, remove the *scraper*—it looks like a washer with wings—and rubber wiper. The retainer and seal come out with a bit more difficulty. With a heavy towel positioned to cushion the shaft's impact and the cylinder pointing away from people, use an air nozzle inserted into the inlet opening to apply pressure to push out the shaft. The shaft will bring the retainer and seal with it, *and may exit with explosive force! Be very careful!*

To rebuild the power-steering unit, consisting of the control valve and power cylinder, you must purchase four parts kits. As of this writing they are available from Ford. These kits are:

ITEM	PART NUMBER
control-valve kit	C3AZ-3A650-B
power-cylinder kit	C1AZ-3A764-A
p/c end-bushing kit	C1AZ-3C59-A
ball-stud kit	C2AZ-3A533-A

Inspect Power Cylinder—Before you assemble the power cylinder, make some checks first. Inspect the center of the shaft for wear by running your fingers over its surface. If you feel any ridges or grooves, don't attempt to repair it; replace the power-cylinder assembly.

Assemble Power Cylinder—The first piece onto the shaft is the new seal. This little rubber ring has a tiny coil spring inside it. After oiling the seal, install it on the shaft with the open side toward the cylinder. Follow the seal with the insert, and then the second seal. The lip on the second seal also faces towards the cylinder. Next on is the retainer with the narrow edge against the seal. Push these in and seat them as fully as possible. Finally, with some more oil, add the rubber wiper and scraper. These parts must now be seated below the retaining-ring groove.

Notice in the photos that Jerry uses a deep-well socket and hollow gear over the shaft that fits snugly over the scraper. With some downward pressure, the parts will seat. Once seated, the retaining ring can be replaced. This finishes the assembly of the power cylinder.

RECONDITION CONTROL VALVE

Disassemble Valve Housing—Now that you've had a bit of practice with the power cylinder, try your hand at reconditioning something more complex—the control valve. To make it as simple as possible, I've broken it down into two sections: the housing and the sleeve.

Begin by removing the housing from the

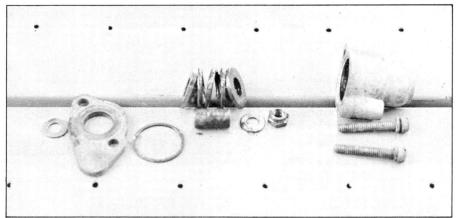

Parts you've removed from reaction-valve housing. All that remains are two bolts that hold housing to the sleeve.

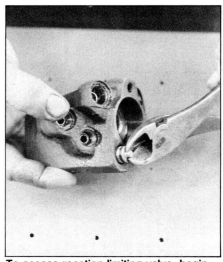

To access reaction limiting valve, begin by removing spring plug. Repair kit includes new O-rings for spring plugs.

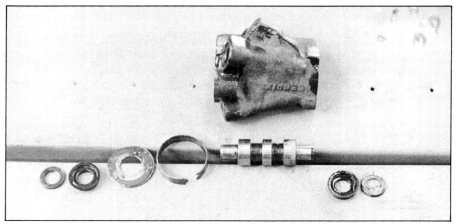

Reaction-valve spool and parts: New seals are in repair kit.

sleeve. To do this, remove the two bolts holding the spring cap to the housing. The spring cap is on the end opposite the sleeve.

Now that the spring cap is off, you have access to the adjusting nut and spring. Remove this nut and washer and slide off the spring seat, spring, spacer and bottom spring seat. Now, remove the adapter and its seal and washer.

To separate the housing and sleeve, remove the two bolts holding the two together. Now, remove the valve spool, seal and bushing. Turn the housing around and remove the spacer and washer from the end that was attached to the sleeve. Also remove the seal and bushing from this end. If your parts look like those in the above photos, you've taken the reaction valve

apart correctly this far.

To finish the job, remove the spring plug, spring and washer. The reaction-limiting valve can be removed. This should be followed by removing the reaction-spring plug and reaction spring from the other end of the housing.

The valve housing should now be completely disassembled. Examine the parts for rust, corrosion and pitting. If you don't find any problems, reassemble the unit using new seals and O-rings. If you find any surface rust, remove it with fine emery cloth. To give the cylinder bore the same treatment, wrap the cloth around itself so it fits snugly in the bore. Move it back and forth while you rotate the cloth. If you prefer, use a bead hone to refinish the bore.

If any serious pitting or rust is discovered, replace the unit. It would be senseless to replace the seals only to have fluid leak due to pitted or rusted areas. How much pitting is too much? Unfortunately, that's a judgement call. If you're unsure, take it in to your local mechanic and get his opinion.

Assemble Valve Housing—If you used emery cloth or a hone, flush the parts with solvent to remove any grit. Dry them and coat all parts, particularly the bores, with automatic transmission fluid (ATF).

Open up the control-valve kit and remove the two small O-rings. These will replace those that go on the reaction-valve-spring plugs. After replacing the O-rings, lubricate them with ATF and slide the reaction valve into its bore. Replace the washer at each end followed by the reaction springs. Slide the spring plugs back in, being careful not to roll off the O-ring. Now, turn your attention to the valve spool.

Look closely at the valve spool. Within one of the grooves—there are three *lands* and two large *grooves*—you'll see a narrow groove. This narrow groove must face the *small* end of the valve housing. If the valve spool is installed backward, the car will not steer properly.

After lubricating with ATF, slide a new

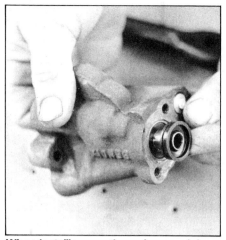
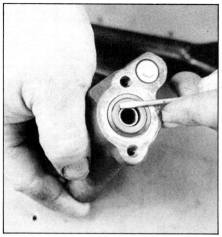

When installing reaction-valve spool, be sure seal faces open-side out so that seal bushing will fit in its seat. Press in bushing firmly.

Sleeve end of housing also gets a seal and seal bushing. Jerry shows how seal lips over edge of valve-spool bore.

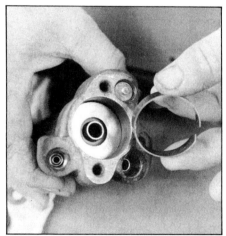

Replace washer and spacer and rebuild job is finished. Washer is shown in place.

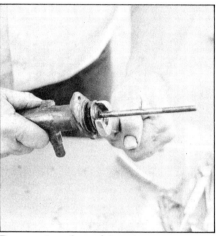

Remove travel-regulator stop as you would any nut.

seal over the end of the valve spool, open side out. Be sure it sits flush with the face of the valve. Next, insert a seal bushing and wedge it in tightly between the spool and seal. Turn the housing around and repeat the operation at the other end. Finish the housing rebuild by inserting the big washer and spacer in the large end. It's the end with the two tube seats.

Disassemble Valve Sleeve—Begin disassembly of the sleeve by pulling the head of the valve-spool bolt tightly against the travel-regulator stop. This will open the unit so that you can remove the stop pin. In the illustration, the stop pin is pushed out with a small screwdriver. If more force is required, use a drift punch and small hammer.

Begin by removing the rubber boot and clamp from the valve sleeve. Then, turn the travel-regulator stop counterclockwise in the valve sleeve to remove the stop from the sleeve. The remaining parts will slide out. Remove the valve-spool bolt, spacer and rubber washer. Slide the bumper, spring and ball-stud socket from the sleeve. Remove the grease fitting and the sleeve is fully disassembled.

Assemble Valve Sleeve—Coat the inside of the sleeve, ball-stud socket (inside and out), and ball stud with white lithium grease.

Insert one of the ball-stud-seat halves, flat end first, into the ball-stud socket. Then insert the threaded end of the ball stud into the socket. Slide the socket into the control-valve sleeve and pull the threaded end of the ball stud out through the slot in the sleeve. Place the other ball-stud-seat halves against the ball stud.

Place the bumper into the spring and insert the two, as a unit, into the sleeve, resting against the ball-stud seat. Insert the valve-spool bolt through the travel regulator and thread this into the sleeve, tightening it securely. Back the travel regulator off just enough to align the nearest hole in the stop with the slot in the ball-stud socket. Install the stop pin in the ball-stud socket, travel-regulator stop and valve-spool bolt. Pull out on the spool bolt and install the rubber seal. Complete the assembly by installing the rubber boot and clamp. Check the ball stud for free movement by sliding it back and forth.

Assemble Control Valve—You can now join the control-valve housing and sleeve, adjust the unit, and assemble the power cylinder to the center link. Start by sliding the housing over the spool bolt and align the bolt holes. Run in the bolts and torque

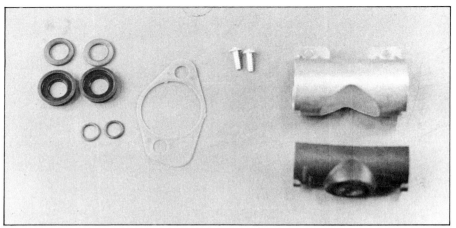

Control-valve rebuild-kit parts.

Insert bumper and spring after installing other ball-stud seat.

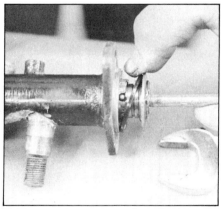

Rubber washer goes on first. Next, seat travel-regulator stop, align holes and insert stop pin.

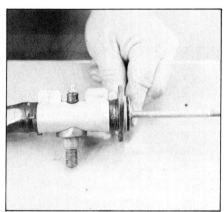

Complete reaction unit with new dust boot and clamp in place.

This unit had two washers. Yours may only have one. Install spring with its two seats. Lock them with a washer and nut. Don't forget to adjust nut according to directions in text. Finish job by installing spring cap.

them about 15 ft-lb. Place the adapter over the other end of the housing followed by two washers. Note: If your unit only had one washer, only replace one washer. Washer count varies from one unit to the next.

Install the spring seats and spring with the spacer in the center and slide this assembly over the spool bolt. Finally, clamp it together with a washer and nut. You must now center the valve spool. To do this, tighten the nut at the end of the spool bolt—about 90 in-lbs—then back off about 1/4 turn. The final test will come when you drive the car. If the car turns to the left easier than it does to the right,

loosen the nut another 1/2 turn. If a right turn is easier, tighten the nut about a 1/2 turn.

If your car came equipped with a six-cylinder engine, centering the valve spool is a little different. With the ball stud in a straight vertical position, measure the distance from it to the center of the first bolt hole—with a bolt in it. This distance should be 4-1/8—4-3/16 in. If it's not, adjust the distance by rotating the center link one way or the other.

Complete the assembly of the control valve by bolting the spring cap to the housing. Don't forget the seal between the cap and housing.

The control valve and power cylinder can now be connected. Begin by bolting the power cylinder to the center steering link. Be sure to install the cotter pin. Next comes the dust boot. Install it over the power-cylinder shaft. Check the hoses. If they're cracked or show the least amount of wear and tear, replace them. Install them as shown in the photos, following page.

From the end-bushing kit, install new bushings and spacers with the "frame" bracket on the end of the shaft. If all hoses are installed, you can now install the control valve-and-cylinder assembly in the car.

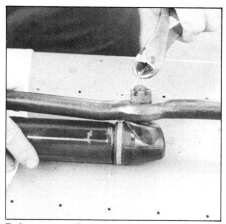

Bolt power cylinder to center steering link. When tightening, align nut castellations with hole in stud and install cotter key.

Install new rubber bushings at power-cylinder ram and mounting bracket.

Install Control Valve & Power-Steering Cylinder—Working from under the car, install the idler arm. Torque the nuts/bolts 30—40 ft-lb. At the other side, install the steering gear. Torque its bolts 50—65 ft-lb. Fit the center link to the idler arm. Secure it by threading on the nut finger tight. Lift the other end up and slide the ball stud into the Pitman arm. Secure it with its nut, again finger tight. Position the power-cylinder bracket against the underside of the "frame" and secure it with its three bolts. The power-assist unit is now in place.

Connect both tie rods to the center link and secure them with their corresponding nuts. Torque all nuts to about 60 ft-lb, then back each off until a cotter pin can be installed. This puts the power-steering unit back in place. Once you connect the lines from the power-steering pump, the power-steering system installation is complete. Final adjustments will be made when the car is complete and ready for a front-end alignment.

REBUILD POWER-STEERING PUMP

The third major power-steering component is the pump. You can rebuild this unit if need be. How do you determine whether or not it needs rebuilt or replaced? If it leaked, tear it down and reassemble it with a new seal kit. If it made noise, replace it. If the pump didn't leak or make noise, clean it up, paint and reinstall it. Always

Coax off power-steering-pump reservoir from pump body with a wood block and hammer.

install *new* hoses that go to the power cylinder, unless they were recently replaced. If you reuse the hoses, clean them with waterless hand cleaner.

A word of warning if you decide to tear into the pump: It is a relatively complex component with a great number of parts. Some of these parts are spring-loaded. If any spring-loaded parts get away from you, they can be reinstalled. However, it's difficult . . . even for a professional. So be very careful as you disassemble the pump. Remove the parts with care, laying them out in their order of removal with one side or the other up in a consistent manner.

Disassemble Pump—Start by draining the pump. Remove the cap and pour out all of the fluid. Turn the pump over and pour out what you can from the reservoir tube. Once again, rotate the pump and pour whatever is remaining from the filler tube. Do this operation two or three times. When you think you've removed as much fluid as as possible, begin the disassembly by removing the drive pulley. When done, remove the adapter nut and identification plate.

Clamp the pump in a vice, gripping a cover boss. Hold a block of wood against the edge of the reservoir and tap around it with a hammer. This will separate the reservoir from the cover. The pump housing is retained to the cover by three bolts and a stud. These should be removed a turn or two at a time as there is spring-loading between the cover and pump housing.

When the last bolt is out, lift the pump housing from the cover, removing the large O-ring and gasket. The parts that remain on

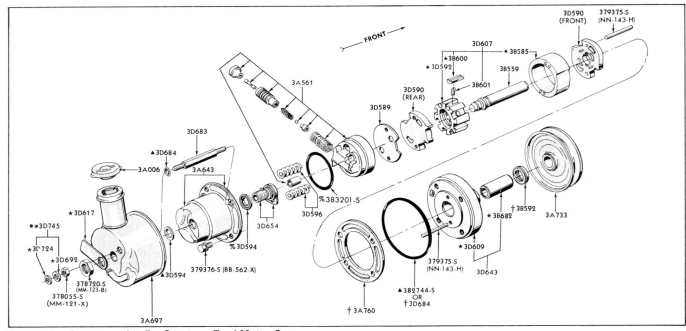

Power-steering pump details. Courtesy Ford Motor Company.

Remove bolts and stud a turn or so at a time because pump housing is loaded by two springs. After bolts and stud have been removed, pump separates into two sections.

Be careful removing cam ring and rotor. Otherwise, you'll end up with all of these pieces on the floor. If this happens, pump should go back together like this. Consult text for assembling pump.

the cover are the spring-loaded ones that I cautioned you about. So proceed with care. Remove the pressure plate to expose the rotor and cam ring.

The rotor and cam ring are the most complex part of the pump. Look carefully and you'll see what I mean. There is an outside ring called the *cam ring*. Within the cam ring is the rotor. Between each fin of

the rotor is a *slipper* held in position by a spring. If, when unloading this unit from the cover, the rotor should separate from the cam, pieces will fly everywhere. To prevent this, lift the unit by the cam ring and carefully slide it off the dowels. Set it aside, face up, the way it came from the cover.

If you're unfortunate and everything

comes apart, here's how to put it back together. Place the cam ring on a flat surface with the elongated portion of the aligning hole face down. On cam rings that have a positioning arrow, face the arrow down. Place a 1/4—5/16-in.-thick spacer inside the cam ring and set the rotor on the spacer, stepped-bore side up.

To place the slippers, insert the blade of

Parts that install in pump housing. Pressure plate at right is in correct relationship to other parts.

Lubricate O-ring seal before installing reservoir on pump body. Use white lithium grease or Vaseline.

a very small screwdriver into the end coil of a spring; insert and compress the spring with the screwdriver. Slide a slipper into place with *its projection facing the direction of rotation*—counterclockwise in this position. Repeat the procedure for the remaining slippers. Refer to the photo at left for how the slippers are positioned in relation to the rotor.

If you successfully removed the cam ring and rotor, remove the pump shaft, followed by the thrust plate. You must now remove the shaft seal. I use a cape chisel to collapse one side of the seal and pry it out.

To remove the manifold assembly from the pump housing, push down on the discharge adapter. This usually works. Sometimes, however, it sticks. In this case, place a shop towel or wood block on the floor. Hold the housing, open end down, and strike it on the towel or wood block. This will jar loose the manifold assembly.

What will come out are the end plate, manifold, two springs and the discharge adapter. In the illustration, I've included the pressure plate to show the relationship of the parts. This completes the pump disassembly. As usual, clean everything and paint the exterior parts—black in this case.

Inspect Pump—If the pump was noisy, check all the friction surfaces for scoring and excessive wear. If you find any such damage, replace the pump. New seals and O-rings won't correct the noise, only a

leak. And you can't get the parts necessary to rebuild the pump. Nor can you get a new pump, only a remanufactured one. However, if all parts look OK, continue on.

Assemble Pump—From the rebuild kit, install the O-ring for the discharge adapter. Insert the discharge adapter into the pump housing. To keep the adapter in place while you replace the manifold and springs, tie or twist a piece of wire around it, or even wrap a rubber band around it. Anything that prevents it from falling out when you turn the housing right-side up.

Install a new seal in the cover. To do this, lubricate the outside of the seal and the inside of the seal bore. Set the seal in place, rubber lip toward the outside. Place a 7/8-in. socket over the seal and gently tap on the socket with a hammer, seating the seal flush with the edge of the seal bore.

Turn the cover over. Note that there are two dowels. One unscrews; the other is permanently installed. The removable dowel is called the *aligning pin*. The thrust plate, cam ring and pressure plate each have one elongated hole. This hole goes over the aligning pin, keeping everything in the correct order.

With this in mind, lubricate the seal with white lithium grease or Vaseline. Clamp the cover into a vise by one of the bosses. Mount the thrust plate, install the pump shaft, then the cam ring and rotor, finishing off with the pressure plate, worn-side

down. Place the new O-ring on the cover.

Align the notch in the end plate with the matching notch in the pressure plate and install it. Then, align the notch in the manifold with the notch in the end plate and install the manifold. Place the O-ring over the groove in the manifold and thoroughly lubricate it with ATF. Place the two springs in position.

Position the housing over the cover with the bypass holes aligned and the gasket between them. Replace the stud in the hole closest to the discharge adapter. Then replace the three bolts. Tighten the bolts and stud evenly, being careful not to cut the manifold O-ring with the inlet slot that's in the side of the housing.

Install the new gasket over the adapter and a new copper washer over the stud. Lubricate the O-ring on the cover, align the reservoir with the adapter and stud and set it into position. It should go on with hand pressure. If not, a *little* nudge with the a wood block and hammer will do the job.

To finish assembling the pump, install the pulley, page 124.

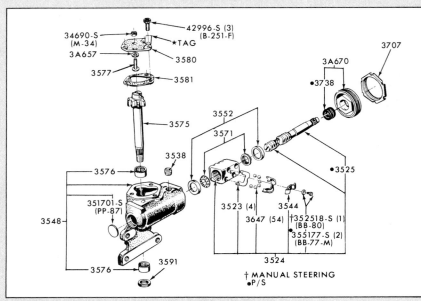

Steering-gear details. Courtesy Ford Motor Company.

How To Rebuild Your Mustang Steering Box

By Steve Hendrickson, courtesy of *Mustang Illustrated* Magazine

Next time you're out in a junkyard, try finding a complete Mustang of the 1965—73 vintage. Mustang restorers have been scavenging parts from boneyard 'Stangs for years, and many parts are almost impossible to come by. Many body and interior pieces for the pony cars are being reproduced, but that doesn't do you any good if you need some of the more choice chassis pieces. Although many of the Mustang's underpinnings were shared with other FoMoCo products—Falcon, Comet, Fairlane, Maverick, Cougar, Montego and Torino, to name a few—some parts are getting pretty scarce.

One Mustang part that's particularly difficult to find in good condition is the 1965—73 steering box. By now, these boxes are around 20 years old, and most junkyard boxes are in sorry shape. Furthermore, many of these boxes on the street have seen thousands of miles, and probably need to be replaced or rebuilt. You can try to find a rebuilt one from a parts store, but it isn't easy. We called around to several in our area, and only one said they *might* be able to get one, but it would cost about $225. Everybody else said, "Sorry, not available."

The next step was to call several Ford dealerships. All of them said that rebuilt boxes weren't available, and all but one said that the rebuilt parts are all obsolete. That one exception was Jim Bremner, who works at Walker-Buerge Ford in Downey, California.

According to Jim, all of the bearings and seals necessary to rebuild a Mustang box are available. The only obsolete parts are the sector and worm shafts, and the cover gasket. The two shafts are tough enough that they'll be good in most boxes, and the gasket you can make yourself. In other words, if you need to rebuild your Mustang box, you're in business.

If you have an "in" at a Ford dealership, you can order the parts through them. Most dealers don't even want to take the time to mess with them, however, because they don't make money on ordering obsolete parts. One source that's guaranteed to cooperate is Larry's Thunderbird and Mustang Parts (511 S. Raymond Ave., Dept. MI, Fullerton, CA 92631; 714/871-6432). They have all the parts listed below in stock.

First, determine if your box has a 1-in. or 1-1/8-in. sector shaft, as they require different bushings. At right are listed the parts you'll need:

Pre-'67 Mustang steering box (left) had integral steering-column shaft. Box at right is example of '67-and-later unit. Rebuild procedures for each is essentially the same.

Check to see if 1- or 1-1/8-in. sector shaft is used before ordering parts; they require different bushings.

After centering input shaft in its full right and left travel, remove locknut that's concentric with input shaft. This will free the worm-bearing preload adjuster.

Pull out worm-and-rack assembly, clean it thoroughly and store assembly in a clean, dry place. A plastic bag will keep it clean and dry.

Loosen sector-shaft adjusting bolt (arrow) and remove side-cover attaching bolts. Treat gasket with care; it's an obsolete part.

You should now be able to push out sector shaft. If you don't have replacement needle bearings, do this over a surface that will catch old bearings as they fall out.

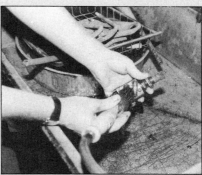

Clean all steering-gear pieces and inspect them for cracks and excess wear. Check housing, too.

1. Steering-gear worm-bearing cups (2) upper and lower—C2AZ-3552-A
2. Steering-gear worm thrust bearings, (2)—C2AZ-3571-A
3. Sector-shaft seal (1)—C702-3591-A
4. Steering-gear sector-shaft bushings (2);
   For 1-in. shafts—
   upper—C4DZ-3576-A
   lower—C3DZ-3576-A
   For 1-1/8-in. shafts—
   upper—C4DZ-3576-B
   lower—C1DZ-3576-B
   These parts run just over $50. We got a

Back at the housing, knock out plug from bottom of housing. This will allow you to remove lower worm-bearing cup.

STEERING BOX, continued

Upper worm-bearing cup is installed in worm-bearing preload adjuster. Remove it and carefully tap in new one. It must go in straight.

Now for the sector-shaft bushings: Remove old ones and press in new ones, taking care not to dislodge needle bearings. As shown, flanged bushing goes on inside.

Carefully tap in bottom bushing until it's flush with edge of its bore.

New seal goes in next. Again, take care not to dislodge needle bearings.

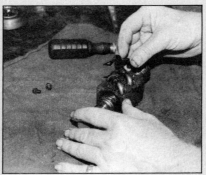

Now, inspect worm-and-rack assembly, clean it with solvent, and rotate it back and forth, checking for roughness.

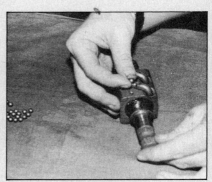

Clean innards, checking each of the 62 ball bearings for pitting and flat spots. Inpect worm shaft for wear and pitting. If shaft is bad, replace steering gear. If all is OK, lubricate shaft and center the rack on it.

used box from a local pull-your-own parts junkyard for $18, so we'll have a good, rebuilt box for about $70. Some sources sell "good" used boxes for that much, and if you consider the cost and availability of a rebuilt unit, doing it yourself looks like a pretty good deal.

Once you have the parts, rebuilding the box is a cinch. First, remove the lock ring from around the input shaft, and then take out the large "pipe fitting" plug. This will allow you to remove the rack-and-worm assembly (it's easiest to pull if you center the steering box first). Clean it up and set it aside for later.

Next, loosen the sector adjusting-screw locknut, remove the three bolts that hold on

the top cover, and pull it off. You can now push the sector shaft out by just pressing it down on the bench. When you do this, chances are needle bearings from the sector bushings will probably go everywhere. If you have the proper replacements, don't worry. Clean the case and sector shaft thoroughly, and inspect for cracks or excessive wear. There will appear to be deep scratches on the sector gear and its mating surfaces on the recirculating block—this is normal and caused by the manufacturing process. If the sector shaft or the worm shaft are worn out, you'll have to find another box—most of the time they're good, though.

Now that everything is out of the housing,

remove the old worm-bearing cups and install new ones, one in the housing, the other in the end cap. New sector bushings are next. Push the old ones out, and gently tap the new ones in. Note that these are different; the one with the lip goes on the inside of the box. Also, be careful not to dislodge any of the new needle bearings. The last step on the housing is to install a new sector-shaft seal.

Now turn your attention to the rack-and-worm assembly. Clean the old grease off, and gently roll the rack back and forth on the worm. If there's any resistance or roughness, you'd better pull it apart for inspection. If everything feels smooth, leave it alone and skip on down to the reassembly.

To check the rack-and-worm assembly, first remove the recirculating tubes, then turn the rack upside-down and turn it from side to side. All of the bearings should fall out, so be prepared to catch them in a suitable container. Once all the bearings are out, you can remove the rack from the worm. Check for wear on the worm shaft, and inspect the bearings for scratches for flat spots. If you need new bearings, mike a few to find the diameter, and get replacements from a bearing supply company. If you don't have a micrometer, just take them down to the bearing house and let them figure it out.

To reassemble the rack-and-worm assembly, lube it with grease, install the rack in the middle of the worm, insert the recirculating tubes, and drop 31 bearings into each passage. Don't turn the worm shaft while inserting the bearings, or they'll get into the wrong passage and eventually fall out. Once all bearings are in, install the cap and and screws. Now, turn the rack from end to end on the worm gear, to see if any bearings fall out. If they do, you have to disassemble the unit and start over.

Once the rack-and-worm assembly is done, you can reassemble the entire box. First, pack the worm bearings and install them in their cups. Then, put the rack-and-worm assembly in, and thread in the worm-bearing adjuster cap on the end. Tighten it to 4—5 in-lb, and install the locknut. If you have done this before, you can probably do it by "feel;" if not, you should beg, borrow or rent the appropriate torque wrench. The worm shaft should be tight enough to turn freely. The last step is to reinstall the plug on the other end of the worm shaft. Just tap it in with a hammer and punch.

Make certain the rack is in the middle of the worm shaft and slide it in. If you were able to save the original cover gasket, use it; otherwise you'll have to make your own from a suitable grease-proof gasket material, available at most parts stores. Install the cover, loosely, with the three bolts. Make sure there's some lash between the shaft and the adjusting screw before you tighten the cover bolts. When you get the box installed on your car, set the lash properly, page 69.

The last step is to remove the grease plug and fill the box with a generous amount of grease. Ford recommends Ford gear lube, part number C3AZ-19578-A, but a high-quality, non-fibrous disc-brake bearing grease is fine. Now, paint it the color of your choice, install it on your car, and enjoy easy, safe steering!

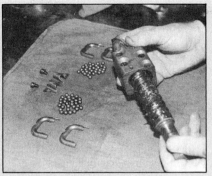

Install bearing tubes and 31 ball bearings in each side. Don't work rack back and forth until you have all 62 bearings in place. Then, rotate rack back and forth to check that no bearings fall out. If they do, tear assembly down and start over.

Grease sector shaft and insert in housing. Use Ford's gear lube C3AZ-19578-A or high-temperature disc-brake grease.

If you couldn't salvage old cover gasket, make a new one from similar gasket material. With adjusting bolt backed out, install cover and torque cover bolts 15—22 ft-lb. Adjust sector-shaft lash after you've installed steering gear on car.

Pack new worm bearing, install them in their cups, and insert worm-and-rack assembly in housing. Input shaft should turn freely, with no end play. Install lock ring and tap in bottom plug. Finally, remove grease plug and pack gear with specified grease.

6
Brakes

Brake components available from Stainless Steel Brake Corporation range from disc-brake rotors to pedal pads. Shown is their disc-brake conversion for '65—67 Mustangs. Photo courtesy Stainless Steel Brake Corporation.

Sheet-metal nut lock takes place of early-style castellated nut. Both use cotter pins.

Mustang brakes, like the rest of the car, are simple and straightforward. This makes restoring the brakes relatively easy. However, don't let this simplicity lull you into complacency. The system must still be given considerable care and attention. Should the brakes fail, a severe accident could follow.

When rebuilding brakes, pay close attention to detail. Check and double-check as you move along. Test the system whenever possible. When you reach the test-drive phase, try the brakes several times as you move along at a crawl. If anything seems wrong, check it immediately and find and correct the problem. With these safety pre-

cautions well in mind, turn your attention to rebuilding the brake system.

DRUM BRAKES (FRONT)

Remove Drums—Loosen the wheel lug nuts slightly. Then jack up the car and place jack stands under the front "frame rail" and rear torque box—it's the reinforced sheet-metal panel adjacent to and outboard of the front eye of each rear leaf spring. Finish removing the lug nuts, then pull off the wheels and tires from the front drums.

To remove the drum from the spindle,

Use K-D or similar-style brake tool to remove shoe-retracting springs. You'll use it later when it's time to assemble brakes.

remove the dust cap by either prying it from the hub with a screwdriver or pair of water-pump pliers. Clip off or straighten the ends of the cotter pin and remove it from the nut lock. Then remove the flat washer and adjusting nut. Wiggle the drum to push out the outer wheel bearing or stick your finger in there and pull it out. The drum should now slide off the spindle.

If there's any resistance on the part of the drum, give it a few good whacks with a wood or leather mallet. This should free it so you can pull off the drum. Sometimes, however, the shoes will have worn into the drum, leaving an unworn edge or rust will build up on this edge. The result is the drum will hang up on the the shoes and not let the drum slide off. If this is the case, you'll have to adjust the shoes away from the drum. Whatever you do, do not try to pry off the drum. Instead, back off the brake shoes.

At the bottom of the backing, or carrier, plate is an access slot for the brake-adjusting mechanism. If there's a rubber plug in this slot, remove it. Stick a screwdriver in the hole and lift up on the brake self-adjusting lever. While holding up on the lever, insert a brake-adjusting tool and rotate the adjusting-screw star wheel. Rotate the wheel *down* several turns by lifting *up* on the handle to pull the shoes away from the drum. Again, you may have to rap on the drum a few times. However, if the drum still hangs up on the shoes, back them off a little more and try again. Don't force the drum off.

Remove Shoes—K-D Manufacturing

makes a brake tool that you should have in your tool box. It makes removing and replacing the shoe retracting springs relatively simple compared to the wrestling match you would have trying to use pliers and a screwdriver.

Using your brake tool, remove both retracting springs from the anchor pin and then from the shoes. In the following order, remove the rest of the brake mechanism: shoe-retracting assist spring, adjuster spring, adjuster cable and adjuster, cable guide, brake-shoe hold-down springs, adjusting-screw assembly and finally, the brake shoes. Everything should be off of the backing plate now except the wheel cylinder.

The wheel cylinder is removed by first disconnecting the flexible brake line. Use a flare-nut wrench here to avoid rounding off the flats on the union. Presoak the connection with rust penetrant if the nut is corroded. Two bolts retain the wheel cylinder to the backing plate. Remove these and the wheel cylinder will come off. If you wish to remove the backing plate, remove the four bolts that hold it to the spindle. Notice that one of these bolts is longer than the other three. It will be obvious where it goes when it's time to reassemble the brakes; just don't try to put the long bolt in the wrong hole.

Backing plate with only wheel cylinder attached.

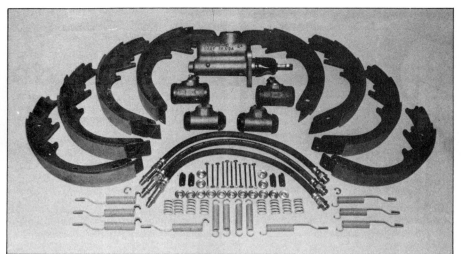

Complete drum-brake restoration kit includes everything you'll need, including springs, shoe retainers and plugs. Photo courtesy Kanter Auto Products.

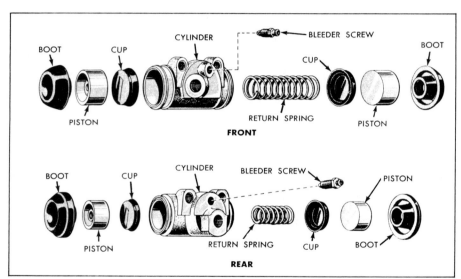

Wheel-cylinder details. Courtesy Ford Motor Company.

REPAIR & RECONDITION FRONT BRAKES

Inspect Drums—Carefully examine each drum. Check for grooving on the friction surface. If there are any deep grooves, have *both* drums turned. Or, if the friction surface is flat, but there is a large unworn edge at the outside edge of the drum, the drum is probably worn badly and should be turned or replaced. Again, have both done.

If you suspect the drum is worn excessively, warped or out-of-round, take it to a brake shop and have it measured with a brake micrometer. Then you'll know for sure.

Inspect Shoes—In the same fashion, the brake shoes should be replaced unless they are practically new. Technically speaking, they may be reused as long as there's 1/16 in. or more material above the rivets. But

this is a restoration, not a "patch job." Regardless of the amount of wear, replace the shoes if they are wet with oil, brake fluid or any foreign matter. However, if wear is minor and the shoes are relatively clean, wash them in denatured alcohol. After the alcohol evaporates, scuff the shoes with #80-grit sandpaper to remove any glazing.

If you want the best possible brake job, purchase premium shoes and have them *arc-ground* to fit the drums. This will ensure maximum shoe-to-drum contact, giving proper braking on your first stop. It also reduces the chance of uneven shoe or drum wear.

Inspect Wheel Cylinders—Wheel cylinders should also be replaced. The price difference between new cylinders and rebuilding your old ones is minimal. The chance is high rebuilt cylinders will leak, especially if there was serious pitting in the bores to begin with. However, if you wish to rebuild the wheel cylinders, follow the procedure for rebuilding a master cylinder, page 94. For the ultimate wheel cylinder, Stainless Steel Brake Corporation sleeves the cylinder with a stainless-steel bushing and fits a brass piston.

Replace Springs & Adjusters—After years of service, the shoe-retracting springs fatigue and the automatic shoe-adjuster mechanism wears until neither do their jobs very well. This is particularly true of the adjuster. So, to restore your brakes to top performance, replace these components. Most good auto-parts stores can supply all the components needed.

Wheel cylinder at left is front wheel cylinder as indicated by its larger piston; rear cylinder is at right.

Bird's eye view of how four wheel-cylinder pistons should be installed. Front is at bottom; rear at top.

White spots around edge of backing plate are dabs of white lithium grease.

Restore Backing Plates—Wash the backing plates in solvent to remove grease, followed by a bath in denatured alcohol—to remove any brake fluid. Aerosol brake cleaner does a good job, too. Remove old paint with paint stripper or beadblasting. Sandblasting works, but pits the metal excessively. This makes it hard to restore the original high-gloss finish. When clean and free of old finish or rust, paint the backing plates gloss black.

If you want less work and a nice, durable finish, have them powder-coated. The powder-coating shop will do all of the work, including stripping, and return to you what looks like a new backing plate. This procedure lasts much longer than conventional paint, but is not original. So beware if you're headed for the show circuit.

ASSEMBLE FRONT BRAKES
Install Wheel Cylinders—Begin assembling the front drum brakes by mounting the wheel cylinder to the backing plate. If you're working on the rear brakes along with the fronts, you'll have four wheel cylinders: all different. Here's the rub; three of the four places they fit are wrong! To tell the four apart, look at the piston sizes. The two front-wheel cylinders have larger pistons than the rears. Check this by removing the dust boot from each cylinder and examining the bore. You can see the difference, or measure them with a vernier caliper.
Install Backing Plate—After determining which are the front wheel cylinders, mount them so the brake-line inlet ports are to the

rear of the car. But first determine which side of the car the backing plates go on. Lay the two plates down side-by-side with the anchor pins at the top. Note the location of the adjusting-screw access-hole in relation to the anchor pin. One is to right-of-center in relation to the anchor pin and the other is to left-of-center.

As you face the plates, the one with the access hole to *right-of-center* goes on the driver's (left) side. This puts the adjusting screw and wheel cylinder in the correct location. Now, mount the wheel cylinders and backing plates on the spindles. Torque the wheel-cylinder bolts 5—7 ft-lb (9-in. brakes) 10—20 ft-lb (10-in. brakes). Torque the backing plate-to-spindle bolts/nuts 20—35 ft-lb.

Install Brake Shoes—It's now time to set the new shoes into place. Lay each pair side by side and you'll notice one shoe has about 20% more friction material on it. The smaller one is called the *primary shoe.* It mounts to the backing plate toward the front of the car; the larger, or *secondary shoe,* mounts to the rear.

Before mounting the shoes, lubricate the rub points on the backing plate with white lithium grease. Then, mount the shoes as described above. Retain each to the backing plate with a shoe hold-down pin, washer and spring.

Press the shoe against the backing plate, then insert the pin through the back of the backing plate. Fit the spring and washer

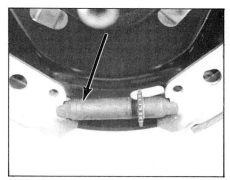

With shoes secured by hold-down springs and pins, adjusting screw should look like this. Notice that nut has only one groove (arrow), indicating left-hand thread; it installs on left side of car. Adjuster wheel goes directly over access port in backing plate.

over the flattened end of the pin, compress the spring and rotate the washer 90° to lock it to the pin.
Install Automatic Adjuster—The adjusting screws have either right- or left-hand threads. If you interchange these assemblies, one side with the other, the automatic adjusters will cause the shoes to retract rather than expand each time the adjuster operates. To prevent this, the *pivot nut* is marked with two grooves to indicate right-hand threads or one groove to indicate left-hand threads. The adjusting screw is marked on its unthreaded end with an R or L for right- or left-hand threads, respectively. The right-hand nut goes on the passenger's (right) side; the left-hand nut (one groove)

Adjuster cable and spring are attached to adjuster lever. Adjuster is then pulled into position on secondary shoe.

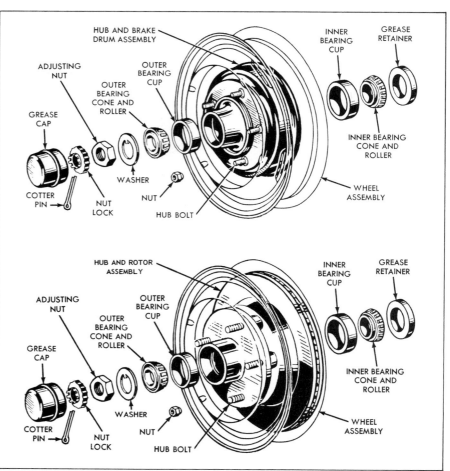

Exploded view of drum-brake/wheel-bearing assembly. Courtesy Ford Motor Company.

goes on the driver's (left) side. Install the adjusting-screw assembly between the shoes at the bottom with the toothed wheel toward the secondary (rear) shoe.

Double-check that the toothed wheel is over the access port in the backing plate. Also check that the adjuster expands the shoes when you rotate the wheel "down." If it isn't or doesn't, the adjuster is backward and/or on the wrong side. Correct the assembly before going any further.

To complete installation of the adjusting mechanism, hook the adjuster cable to the anchor pin. Hook the sheet-metal cable guide into the secondary shoe. Now, mount the shoe-retracting assist spring between the two shoes. This will pull the shoes up to the anchor pin. Next, install the primary and secondary shoe-to-anchor springs.

The adjusting lever is marked with an L or R. In this case the L and R indicate on which side of the vehicle the component installs. For example, the L lever goes to the left (driver's) side and vice versa.

Attach the adjuster cable to the adjuster lever. Attach one end of the adjuster spring to its location on the primary shoe and the other end into the adjuster lever. Pull the adjuster lever into its position on the secondary shoe. If your brake assembly looks like the one in the above photo, you've done the job correctly.

To make sure the automatic adjuster works properly, pull on the cable, then release it. The adjusting lever should move up one tooth on the adjusting screw. When the cable is released, the lever should turn the screw the distance of that one tooth.

Service Wheel Bearings—When you removed the drum, the wheel bearings and grease seal came with it. At the very least,

the wheel bearings and races—Ford calls them *bearing cups*—need to be cleaned and inspected.

The outer bearing may or may not be in the brake drum. If it didn't fall out with the adjusting nut, locknut and washer, fish it out with your finger.

On the back side of the drum, pry out the grease seal with a screwdriver. Take care not to scratch the grease-seal bore with the screwdriver tip. Once the seal is out, the inner bearing will fall out.

The inside of the drum/hub assembly will be full of old grease. Wash out this inner cavity, both bearings, the washer, adjusting nut, locknut and dust cap. Use clean solvent so you don't wash old dirt and metal chips from previous cleaning jobs into the bearings.

If you have compressed air, blow all parts dry. Direct the air between the bear-

ing cage and inner race, blasting old grease out from the rollers. Never use compressed air to rotate the bearing. This makes a neat siren-like noise, but damages the bearing rollers. If done to excess, the bearing could come apart, sending rollers flying with possible injury to bystanders. All you want to do is thoroughly clean the hidden recess in the bearing. Several solvent-and-air cleanings will get the bearings glistening clean.

If compressed air isn't available, do the best job you can with paper towels. Don't use cloth or you'll end up with a bunch of lint in the bearing which contaminates the grease.

Inspect the bearings for wear. The rollers should not show any rust, pitting or wear. There will be a slight burnishing of the rollers—in other words, you'll be able to tell where the rollers meet the inner race, but they should show no signs of metal removal or transfer. If you do see any of these trouble signs, replace the bearing. Bearings usually last the life of the car, so don't expect to find trouble unless the car sat for long periods, or water found its way into the bearings from flooding or an extremely humid environment.

Inside the hub, visually inspect the inner races. They should show no signs of metal transfer, rusting or pitting. If so, replace them. Replace inner races by knocking them out with a drift from the back side. Install the new race with a brass drift from the front side. Use only a brass drift, or you can bet your last dollar the steel drift will slip and gouge the new race. Then you'll have to buy a new one and start over.

Pack the bearings with wheel-bearing grease. I use disc-brake wheel-bearing grease, even on drum brakes, because it can tolerate higher temperatures without breaking down. If you have a wheel-bearing packer, great. If not, pack the bearings by hand.

With clean hands, scoop a generous blob of grease into the one palm. With your other hand, roll the bearing into the grease, pressing the bearing into your palm. You must get pressure between the bearing and palm to drive grease between the rollers and cage. Grease will squish out the top of the bearing when done correctly. After grease squishes from the top of the bearing, rotate the bearing slightly and do the next section. Continue until grease is driven through the entire bearing. Wipe the excess on the outer face of the rollers. There is no need to glob grease on the outside of the bearing. Also, there is no need to fill the cavity between the bearings with grease. This does no good, and will only result in a gooped-up brake later when the grease seal is bombarded by an avalanche of hot grease.

You can wipe a light coat of grease onto the inner races in the hub. This will help protect the bearings and race during the first drive.

After greasing, lay the inner bearing in its race in the hub. Wipe off your hands, and install the new grease seal. Never reuse an old seal. Tap the new seal into position using a hammer and socket around the circumference of the seal to evenly distribute the impacts. The seal must go in straight, not cocked. Continue driving the seal until it bottoms in its bore. Turn the drum over and install the outer bearing in its race.

On the car, make sure the spindle is perfectly clean. Wipe a very light coat of grease on the grease-seal lip, then place the drum/hub combination on the spindle.

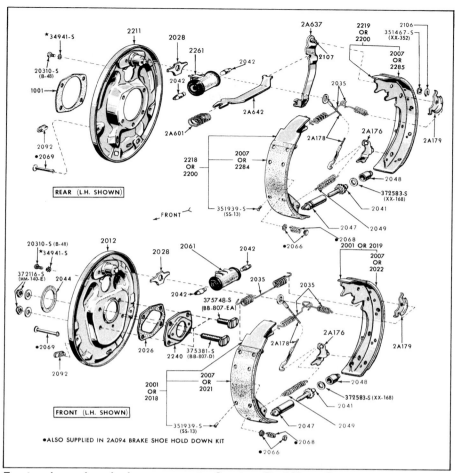

Front and rear drum-brake components. Courtesy Ford Motor Company.

Hold the drum steady, and push it straight on the spindle. Don't rest the grease seal against the spindle as you slide the drum on. This will damage the seal.

With the drum on, place the outer bearing over the spindle and into the hub. Install the washer and adjusting nut. Run the adjusting nut down to just snug. Now rotate the wheel and tighten the adjusting nut to 17—25 ft-lb. It helps to have someone turn the wheel while you tighten the nut. This tightening seats the bearings against their races.

On '64—68 cars, place the locknut over the adjusting nut so the castellations on the locknut are lined up with the cotter-key hole. Now back off both the adjusting and locknuts until the next castellation on the locknut lines up with the cotter-key hole. Install a new cotter key, bending its ends around the locknut.

On '69—70 cars, do not install the locknut yet. Instead, back off the adjusting nut at least a half turn. Then retighten it finger tight, or 10—15 in-lb. Now install the locknut so its castellations line up with the cotter-pin hole, and install a new cotter pin—no nails!

Preliminary Brake Adjustment—The brakes need a preliminary adjustment before the car can be driven, however. This ensures that the wheel cylinders don't stroke too far to push the shoes in contact with the drums and enables the automatic adjusters to work.

Ford specifies a measuring procedure. The brake-drum inner diameter is measured, and the brake shoes are adjusted to just slightly less than that diameter. This is a handy method for Ford dealerships and specialty shops because they have a purpose-built gage which easily and accur-

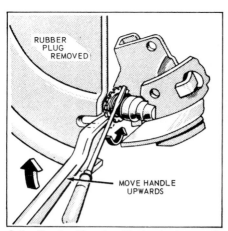

Adjusting tool is raised to back shoes off from drum; lower tool to tighten shoes. Small screwdriver is used to raise lever off star wheel for easier turning.

Insert parking-brake cable through rear backing plate and attach lever assembly to cable end so it looks like this.

ately measures the drum. If you have large calipers, you can get the drums close enough.

But most people don't have calipers of any size, so another method must be used. Reaching through the slot in the backing plate, use a screwdriver to hold the adjusting lever up off the adjusting screw. Now rotate the adjusting screw. Move the adjusting-tool handle—usually another screwdriver—down to tighten. Moving the tool up will loosen the adjustment. Tighten the adjustment until the brake locks the wheel from turning. Now back the adjustment off 10 teeth. The automatic adjusters can take over from here.

To work the auto adjusters, apply the brakes firmly while backing up. After each reverse brake application, drive forward and apply the brakes. The auto adjusters will tighten the shoes during the reverse applications.

DRUM BRAKES (REAR)

Disassembly—If you've completed the front brakes—assuming your car has drums rather than discs at the front—you should have no trouble with the rears. The process is almost the same except the rears incorporate a parking-brake mechanism. Another difference, not related to the brakes, is the rear drums aren't integral with the wheel hubs.

Start by removing the drum. There may be *Tinnerman*—sheet-metal—nuts on

three of the five wheel studs. These are for holding the drum to the axle flange when the wheel is off. If you can't remove these with your fingers, use a pair of pliers to get each Tinnerman started. Then remove them with your fingers.

Once the Tinnermans are off, you should be able to pull off the drum. However, if the drum hangs up because of an unworn or rusted ridge on its inboard edge, back off the adjuster as illustrated in above drawing. Once the drums are off, disassemble the rear brakes following the same procedure outlined for the fronts.

Remove Parking-Brake Cable—After removing everything, you'll find the secondary shoe dangling from the parking-brake cable via the parking-brake lever. To disconnect this cable, compress the spring that's around the end of the brake cable and unhook the lever.

Sometimes, new brake shoes come without the parking-brake lever. Check to make sure yours include this lever. If not, transfer the levers from the old shoes to the new ones. They go on the larger (secondary) shoe in each pair. Simply pop off the retainer from the lever pivot pin with a pair of pliers, transfer the lever to the new shoe, and pop the retainer back on. Using the pliers, pinch the retainer together so it won't come off the stud.

Remove Parking-Brake Cable—The parking-brake-cable housing is retained to the backing plate by a spring-type retainer.

To withdraw the cable from the backing plate, compress the retainer legs below the level of the backing plate shoulder and push the cable housing out. Push on the end of the cable housing as if you were trying to push it out of the backing plate. While you're still pushing, go around the retainer and compress each leg with a screwdriver. When the last leg snaps into the hole, withdraw the cable.

Remove Wheel Cylinders—The only thing left to remove from the backing plates should be the wheel cylinders. Each cylinder is removed like its cousins up front. Disconnect the brake line using a flare-nut wrench, remove the retainer bolts and off comes the wheel cylinder.

Each rear-brake backing plate is retained to the rear-axle-housing flange with the same nuts and bolts that retain each axle. And, the axle flange is between the backing plate and axle housing. So, when you remove these nuts and bolts, the axle must come out before the backing plate can come off.

RECONDITION REAR BRAKES

Everything I said about reconditioning the front brakes applies to the rears. Clean all the parts in denatured alcohol to remove any brake fluid. Paint or powder-coat the backing plates. Also, replace the shoes, wheel cylinders, adjusters and springs and have the drums turned. Not only are you dealing with appearance here, but with durability, performance and safety.

Assuming you've reconditioned and collected all the necessary parts, you can now reassemble the rear brakes.

ASSEMBLE REAR BRAKES

Install Wheel Cylinders—Bolt each wheel cylinder to its backing plate. Like their front counterparts, the brake line enters the cylinder from the rear. Attach the cylinders with two bolts and torque them 5—7 ft-lb (9-in. brakes) or 10—20 ft-lb (10-in. brakes).

Install Backing Plate—Install a gasket first, then fit the backing plate to the axle housing. Install another gasket, then the axle. Tighten the four bolts 30—40 ft-lb on V8 models and 25—35 ft-lb on six-cylinder models. Connect the brake line to the wheel cylinder.

Install Brake Shoes—Pass the parking-brake cable through the backing plate so its spring retainer snaps into engagement. The retainer legs should be sprung out against the inside of the backing plate. Check this by trying to pull it back out.

Compress the spring that's on the end of

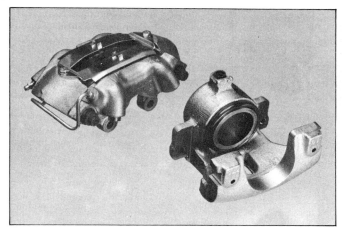

Two basic types of Mustang disc-brake calipers: four-piston at left; single-piston/floating caliper at right. Photo courtesy Stainless Steel Brake Corporation.

New stainless-steel piston compared to typical corroded production disc-brake piston illustrates typical problem with four-piston calipers. Photo courtesy Stainless Steel Brake Corporation.

the parking-brake-cable housing and install the parking-brake lever on the end of the cable. Fit the shoes in place; the big ones (secondary shoes) to the rear and the small ones (primary shoes) to the front. From here on, the procedure is identical as that for installing the front brakes.

Your front and rear drum brakes are now assembled. Do a preliminary adjustment on them and you're finished for now with the rear brakes. For now, let's look at the problems of disc brakes.

FOUR-PISTON DISC BRAKES

Disc brakes used through '67 are the four-piston type. Unlike the later disc brakes, the caliper is fixed, and does not slide on a set of pins. In many ways, this fixed feature makes the early brakes easy to work on.

Disassembly—To remove the caliper, start with the brake hose. Disconnect it at the inner fender, where it passes through the standoff bracket. Use pliers to pull the retaining tab off the bracket. This frees the hose-to-brake line connection so you can get at it. Use only a flare-nut wrench on the fitting or you'll round off the hexes. Plug or tape over the open brake lines to keep out dirt and moisture and to prevent brake fluid from dripping on your face.

Remove the brake hose at the caliper, as well. You should install new hoses, so the sooner you get it out of your way, the better. Aside from that, it's excellent practice to completely remove the hose during any brake work. Twisting the hose, letting the caliper hang from it and even twisting the caliper on and off the hose instead of removing the hose from the caliper will

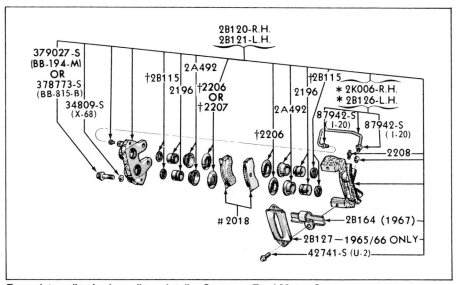

Four-piston disc-brake caliper details. Courtesy Ford Motor Company.

damage it. The damage may not show, but after some time in service, the hose will leak or expand under brake-line pressure.

At the caliper, remove the two large bolts from the inside of the caliper. Leave the two bridge bolts at each end of the caliper alone—they hold the two caliper halves together. Remove the two bolts closer to the center. They hold the caliper to the steering knuckle. When they come out, the caliper will fall free. It's heavy, so be ready.

Take the caliper to the bench and clamp its mounting ears in a vise. Watch how you handle the caliper or brake fluid will spill out. Glycol-based brake fluid makes great paint stripper, so watch it around painted surfaces!

Remove the two small bolts and take off the splash shield. Lift out the two pads, if they haven't fallen out already. Now you can see the pistons inside the caliper. Reach in and remove the dust boots around each piston. A pocket knife or very small screwdriver is a big help here. Lift one lip until you can grasp the boot in your fingers, then peel it off.

The pistons are next. If you are lucky, you might be able to pull them out by hand. Retaining-ring pliers with a 90° head work well if expanded inside the piston's ID. If you are very careful, you can pry the pistons out with screwdrivers held in the dust-boot grooves. Avoid this method as the tools invariably slip and scratch the piston. That means a new piston. The pistons and bores are critical parts which can't tolerate much abuse.

Sometimes, the pistons cock on the way out. Then it is best to push the piston back in and try again. Continuing to tug on a cocked piston is a futile effort. Just knock it back in so it will straighten out and start over. Other times, the pistons are really tight in their bores. Then stick a piece of wood or wadded-up shop towels between the pistons and carefully apply compressed air to the brake hose. This will blow the pistons out of their bores and against the wood. *Make sure your fingers are not in between the pistons and wood!*

And then there are those pistons that are too stuck to come out short of nuclear war. Prying, compressed air and a lot of inventive language won't touch these rusted and corroded examples. Don't bother with them once you've figured they aren't coming out. Their bores are only going to be way past usefulness, so save yourself a lot of time and trouble and exchange them for rebuilt or new calipers.

For those pistons that do come out, pry off any remaining dust boots. Pry the piston seals from the bores. Clean all metal parts in denatured alcohol or brake cleaner, not solvent. Inspect for pitting, rust and corrosion. Clean all seal and boot grooves with paper towels until every trace of dirt is gone. Watch for scratches, nicks and worn areas. All are cause for replacement.

Ford manuals warn against splitting the caliper halves. This is because the caliper hardware is 12-point-head aircraft bolts. If you want to split your calipers to hone the bores, it's OK. Just remember to reuse the exact hardware that came off the caliper—or replace it with high-quality better-than-grade-8 parts.

Some mechanics hesitate before splitting disc-brake calipers because they are so difficult to reseal. This isn't a problem with Mustang four-piston calipers because they use an external pipe for shuttling brake fluid from one side to the other. As long as the transfer pipe is tight, there will be no leaks from splitting the calipers.

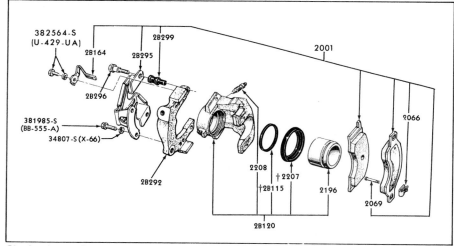

Single-piston, floating-type disc-brake caliper. Courtesy Ford Motor Company.

The transfer pipe can be a problem all its own. Ford used thin material for this pipe, and it won't tolerate much bending. Because splitting the calipers means bending this pipe somewhat, the pipe may need to be replaced if the calipers have been split before. You can make up a new line, preferably out of stainless steel—like professional Mustang brake rebuilders do.

If you want the bulletproof solution, get a set of calipers from Stainless Steel Brake Corporation. As their name implies, these calipers have a stainless-steel lining in the bores and replace Ford's aluminum pistons with solid stainless-steel examples. Rust and corrosion can't get started on stainless steel, so these calipers should last a long time. Stainless Steel Brake Corporation sure hopes so; they guarantee their calipers for the life of the car.

Assembly—Put the caliper in the vise, and set a small dish of clean brake fluid nearby. Dip or roll the pistons in the brake fluid until completely wet. Do the same to the new piston seals. Now install the seals in the bores. Set one side of the seal in the groove, then work the remaining section in by hand.

Install the wet pistons in their bores. Make sure the boot retaining groove faces the caliper center—open section of the piston inward. Gently push each piston straight in. Cocking the piston can roll the piston seal out of its groove. Slow, steady pressure is best. If the piston hangs up, remove it and take a look in the bore. The

seal may have jumped out of its groove, or there could be a foreign object blocking the way. It can take considerable pressure to get the pistons past the seals, so use a lot of brake fluid as lubricant, and keep at it. Top off the caliper with new dust boots, working them fully into their piston grooves. Install new pads and the splash shield.

Disc-brake-rotor service is discussed in the next section on Sliding Caliper Disc Brakes. Follow the instructions given there for rotor removal, inspection and replacement. Wheel-bearing service is covered there, too.

After servicing the rotors, install the calipers back on the car. You may need to lightly pry the pads open when fitting the caliper over the rotor. If you do any prying, keep any eye on the opposite set of pistons. They may come out as you press your side in. If so, use two screwdrivers crossed across the pads and push both in simultaneously. Set the caliper in position and install the two mounting bolts. Torque is 45—60 ft-lb. Connect the flexible brake hose to the pipe, then tap on the clip.

Bleed Brakes—Bleed the brakes after doing any rear-brake work. Brake bleeding is outlined on page 129. If you aren't going to do any other brake work, at least make sure you pump the brakes up. Right now, the pistons are deep in their bores and need a few pumps of the brake pedal to push out against the pads. Don't drive or push the car only to have the brake pedal go to the floor when you need to slow or stop.

To disassemble, sheet-metal stabilizer is first to come off single-piston caliper.

After removing stabilizer, floating caliper and its shoe can be removed from anchor plate.

As an aside, do not use anti-squeal compounds on the brake pads. Rebuilders have found the compound, which sticks the piston and pad together, is responsible for several brake failures. Once the pad and piston are joined together, the pad will pull the piston forward during brake application. This can cock the piston in its bore. When the brakes are released, the piston returns to its at-rest position, pinching the seal. The seal can eventually break, causing a loss of braking and requiring caliper overhaul. If squealing brakes are a problem, use softer pads or piston-to-pad anti-squeal shims.

SLIDING-CALIPER DISC BRAKES

Disassembly—For demonstration purposes I selected a 1969 disc-brake assembly to go through. Notice for assembly that the spindle has been removed from the car. This gives you an opportunity to view things from the backside, a position that is not easy to photograph when on the car. To orient yourself with the photographs, the upper arm is to your top-right as you view the photos. It's been swung around and is being used as a supporting leg for the rotor. Please understand this was done as a convenience only because you don't have to remove the spindle to get to the brake caliper.

Begin removal by unthreading the brake hose. Completely remove the brake hose from the car so you can install new ones when the caliper goes back on. If the hose

was recently changed and you want to reuse it—which you shouldn't—be sure to use two flare-nut wrenches at each fitting. Do not twist the hose or allow the caliper to hang from it. This damages the hose and will cause leaks, pulling brakes or even a complete brake loss. Cap or tape the end of the line to seal out dirt.

Two bolts, which are normally safety wired, retain the caliper anchor plate to the spindle. You can either remove the caliper and anchor plate as a unit, or merely remove the caliper from the anchor plate. If you want to detail the spindle, then it's necessary to remove the anchor plate. But for functional-only brake jobs, just remove the caliper from the anchor plate.

To remove the anchor plate and caliper, snip the safety wire and remove the two bolts. It will take a breaker bar to remove these high-torque fasteners. Once the two bolts are off, the caliper-and-anchor-plate assembly will be loose. Set it on the bench.

To remove just the caliper, remove the four fasteners on the back side of the caliper. The two forward ones are the stabilizer bolts, the two rearward ones are caliper locating pins. They are the pins the caliper slides on during braking. Lift off the caliper and set it on the bench.

To remove the rotor from the spindle, remove the grease cap, cotter pin, locknut and adjusting nut. The rotor will now slide off the spindle. Don't let the outer wheel bearing fall to the ground. Set the rotor and bearings aside until it is time to inspect and restore the parts.

All that's left is the splash shield. For a functional-only brake job, the shield can stay put. But for detailing, the shield must come off. Three bolts hold it to the spindle. Remove them and the shield and its gasket will come off.

If you removed the caliper and anchor plate as a unit, separate them now. Remove the two caliper locating pins and stabilizer bolts from the caliper. These are the four bolt heads visible on the back side of the caliper. The caliper locating pins are actually special bolts, so until they are removed, they look just like two bolt heads. When the bolts are out, the anchor plate and stabilizer will be free. Lift them out and set them aside. The anchor plate is a large, U-shaped casting.

On all brakes, the outboard brake pads are next. Remove the bent sheet-metal clips from the outside of the caliper. This frees the retaining pin the clips attached to, so you can remove them and the outer pad from the caliper. The inner pad is not retained, and will come out of the caliper with a little fiddling.

Last is the piston and seal in the caliper. They are popped out with compressed air applied to the brake fluid passages. But first, place one of the old pads wrapped in a shop towel between the piston and opposite side of the caliper. A block of soft wood works even better if you have some at hand. This protects the piston from dings and scratches when it shoots out.

Using an air nozzle with a rubber tip, place the tip firmly in the hole from which you removed the brake line. Make sure that your fingers are not between the caliper and piston. If they are, they will be crushed when the piston flies out. Now, gently apply air pressure. You want to use the least amount of pressure necessary to pop the piston from its bore. Sometimes seal friction requires considerable pressure before the piston will come free. Then the piston shoots from the bore, slamming into the wood or old brake pad. That's why your fingers must not be in the way.

If the piston won't move, squirt penetrating oil in the brake-line hole and let the assembly soak for awhile. Then reapply air pressure until the piston comes free. If you have to apply penetrating oil, then you can bet the caliper and piston are corrosion-damaged.

Once the piston is out, remove its dust boot. Also reach into the piston bore and pry out the seal. A knife, small screwdriver or, better yet, a plastic seal puller is a big help.

Remove clips from pins so stationary shoe can come off.

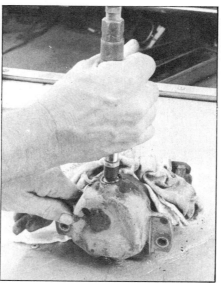

Recommended way to remove piston from caliper. Before applying air pressure, place towel as shown to prevent damage to piston.

These are the two pieces you get in caliper-rebuild kit. O-ring goes into *second* groove inside piston bore. Dust boot fits on piston as shown. Coat all parts with brake fluid before assembling.

RECONDITION & ASSEMBLE DISC BRAKES

Calipers—Inspection and replacement procedures for a brake-caliper bore is similar to that for a drum-brake wheel cylinder. If rusting or pitting of the bore is excessive, the housing should be replaced. Minor surface rust can be removed with emery cloth; minor pitting can be removed by honing.

An alternative to caliper overhaul is caliper exchange. Stainless Steel Brake Corporation reconditions calipers on an exchange basis as well as selling overhauled units outright. Because they coat the bore with stainless steel, their calipers are better than new. Stainless steel resists rust and corrosion much better than the original caliper, so the repair is longer lasting than original. Sure the cost is higher than installing a kit, but the result is better than new. Plus, it means less time and work for you.

You may have noticed I said they *coat* the bore with a stainless-steel treatment, not install a stainless bushing. This is because the Ford calipers have the piston seal mounted in the bores, not on the piston. If a bushing is installed, it must be two-piece to surround the seal. This allows the possibility of having the two pieces shift and cut the seal. To avoid that potential catastrophy, Stainless Steel Brake applies the coating, which cannot harm the seal, yet provides the long-term durability of a bushing.

Even if stainless steel doesn't sound like a good idea, you're better off replacing heavily pitted calipers. By the time deep pits are honed out, the caliper bore will be considerably oversize. So, if the pits are deeper than 0.003 in., the bore will not clean up before getting too large. Replacement—or stainless treatment—are the only cures.

Begin reconditioning the caliper by cleaning all of its parts in denatured alcohol. Clean the piston with very fine emery cloth or steel wool. If there are pits you can't remove or if the chrome plating is worn off, replace the piston. The same applies to the piston bore. Clean it with emery cloth and inspect for pitting. Clean up minor pitting by honing the bore, pages 94 and 95. Although a master-cylinder bore is being honed in that picture, the procedure is the same. Only the hone size is different. Assuming the bores and pistons are OK, you can reassemble the calipers with new O-rings and boots.

Along with new brake pads, get two caliper repair kits. These have the O-rings and dust boots you'll need. Coat The O-ring with brake fluid and insert it into the second grove from the front in the piston bore. Now coat the piston and bore with more brake fluid. Carefully insert the piston into the bore. Check to be sure the O-ring didn't bind, twist or tear. When

you're satisfied all is well, install the dust boot.

Lubricate the dust boot with brake fluid and slide it over the piston, securing the lip of the dust boot in the piston groove. With a small screwdriver, work the other lip of the dust boot into the corresponding groove in the piston bore. The piston should now slide back and forth with the dust boot working like an accordion.

The brake-pad kit should include two sets of pads and two anti-chatter gaskets. The gasket is fixed to the back of the stationary, or outboard, pad. Pull the protective strip from the back of the gasket to expose the adhesive. Press the gasket in place. Do not use anti-squeal compounds on the brake pads.

The new pad and gasket are held in the caliper by the two pins and clips you removed earlier. Replace these pins and secure them with the two clips.

The floating, or inner, pad is retained by two spring clips. Slide this pad into place and fasten the two clips. The caliper is complete and ready to install.

Recondition Rotors—Disc-brake rotors are subject to the same wear and problems as a brake drum. Run your fingernail across both friction surfaces. Can you feel any deep grooves? If so, the rotor should be resurfaced. Do the same if you could you feel any pulsation or vibration through the

After installing O-ring and dust boot, *carefully* work dust-boot edge into first groove in piston bore.

Assembled caliper with shoes in place: Spread caliper wide open so it will fit easily over rotor.

If you don't send rotor out for resurfacing, sand off surface rust and minor imperfections.

brake pedal when you applied the brakes? This could indicate warpage—lateral runout—or rotor-width variation. If you suspect such problems, have the rotors checked at a brake shop. They can cut off—turn—the disc to expose fresh, parallel and constant-thickness material. If the rotors have been turned before, turning them again may cut them undersize. Then, only new rotors will do.

Even if you don't have the rotors resurfaced, get out some coarse emery cloth and remove any surface rust from both sides. Have the *hat*—the section between the rotor and hub—bead-blasted to clean it up. You could do this by wire brushing, but bead blasting is much easier.

Wheel Bearings—Removing, cleaning, inspecting and installing disc-brake wheel bearings basically follows the same procedures given for drum brakes on page 86. There are a few differences besides the bearings being housed in the disc hub and not the drum, however.

First, only wheel-bearing grease labled for use on disc brakes should be used. Disc brakes run hotter than drum brakes, and require high-temperature grease. Using grease not specified for disc brakes can result in wheel-bearing failure.

Also, the inside cavity between the inner and outer bearings should be partially packed with grease. Pack grease into this cavity

only until it reaches the rim the bearing races seat against. Any more is too much.

As always when working with bearings, chase dirt like a Dutch maid, and use non-cloth towels.

Brake Assembly—If you removed the splash shield, replace it now, using a new spindle-to-shield gasket. Tighten the three bolts 9—14 ft-lb.

Install the rotor on the spindle, the outer bearing, washer and adjusting nut. Tighten the nut finger-tight. For now, install the locknut and cotter key without bending the cotter key. You can adjust the wheel bearing after the wheel and tire are mounted. Follow the procedure given on page 86.

If you removed the caliper and anchor plate as a unit, set the assembly in position and thread in the two mounting bolts. The upper bolt—the one facing forward—gets tightened first. Torque it 110—140 ft-lb. The lower bolt gets 55—75 ft-lb.

Use mechanic's wire to secure the bolts. You can get it at any auto-parts house. To wire the bolts, snip off a generous length of wire. Thread it into the drilled bolt head until equal lengths of wire protrude from each side. With your bare hands, twist the wire together. Hold the wire so it forms a 60° V. This will give the proper twist, and look good. Keep twisting, hand over hand, until the twisted portion reaches the spindle. Loop the wire around the spindle, then

Mount completed caliper to spindle. Don't forget to safety-wire bolts.

twist it together some more. After a 1/2-in. length is twisted, snip off the wire and bend back the pigtail so you don't stab yourself with it.

If you dismounted the caliper from the anchor plate, set the caliper in position and install the two caliper locating pins (25—35 ft-lb) and two stabilizer bolts (8—11 ft-lb).

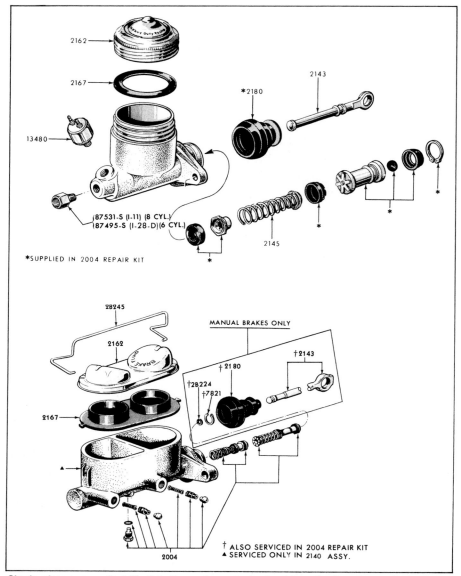

Single-piston, so-called "fruit-jar", master cylinder is at top. Dual-system cylinder is at bottom. Reservoir at rear of dual-system unit is for front brakes and front reservoir is for rears. Courtesy Ford Motor Company.

Diagram labels (top, single-piston)

2162
2167
13480
87531-S (I-11) (8 CYL.)
87495-S (I-28-D) (6 CYL.)
*SUPPLIED IN 2004 REPAIR KIT
*2180
2143
2145

Diagram labels (bottom, dual-system)

2B245
2162
2167
MANUAL BRAKES ONLY
†2143
†2180
†2B224
†7821
2004
† ALSO SERVICED IN 2004 REPAIR KIT
▲ SERVICED ONLY IN 2140 ASSY.

REBUILD MASTER CYLINDER

The master cylinder can be rebuilt if there is only *minor* pitting inside the piston bore. Anything more than that and it should be replaced.

The master cylinder shown in the example is from the 1966 demonstration car. There is also a master cylinder used with a booster which is quite similar. Dual master cylinders were used on all '67-and-later models. Except for the number of parts, there's little difference in working on any of these master cylinders.

Disassemble Master Cylinder—Start master-cylinder disassembly by cleaning its exterior. Remove the reservoir cover and its retainer. Remove the dust boot. There's a new one in the rebuild kit, so don't worry about saving the old one. Next comes the retaining ring. You'll need some internal retaining-ring pliers for removing this. With the retaining ring out, you can remove the piston assembly, primary cup, spring and outlet check valve. If the line fitting is still in the body, remove it.

On dual master cylinders, there are two pistons: The primary piston (it's at the rear and operates the front brakes) and the secondary (it's at the front and operates the rear brakes). Remove the secondary-piston stop bolt (it's under the master cylinder), primary and secondary system spring, valve and tube seat. Also, remove the bleeder screw. The master cylinder should now be stripped down to the bare casting.

Brake Cylinder-Bore Honing—Select a spring-loaded hone such as that shown nearby. These are sold in all auto-parts stores. Select the size that matches the bore you're going to hone. One suitable for the master cylinder or wheel cylinder will be too small for a single-piston disc-brake caliper.

Install the hone in a variable-speed electric or pneumatic drill motor. Use brake fluid to lubricate, and flush metal and abrasives from the bore. Insert the hone into the bore and start it rotating after applying some brake fluid.

As the hone turns, move it back and forth in the bore. *Don't bottom the hone against the end of the bore or pull it out on the return stroke. Either could damage the hone or the bore.*

Continue this operation, adding fresh brake fluid after every few strokes. Stop every five seconds to see how your work is progressing. Continue until all rust and pitting have been removed. When done, thoroughly wash the cylinder with solvent to remove the abrasive and metal filings. *Check all ports and vents to be sure they*

Install a new flexible brake hose between the rigid brake line and the caliper. Use only flare-nut wrenches to avoid rounding off the hexes. Also take care not to damage the new hose by twisting it or letting the caliper hang from it.

Do the other side and your disc-brake rebuild job will be complete. The last thing remaining is to bleed the brakes, page 129. Be sure to bleed the system before driving. At the very least, pump the brake pedal up so you'll have *some* brakes at the first pedal application. If not pumped, the caliper pistons will be recessed deep in their bores, requiring several pumps before they give any stopping action.

are clear. If debris is left in the cylinder, it will fail.

Master-Cylinder Assembly—After assuring yourself the master cylinder is clean, coat the bore and new piston components with brake fluid. Seat the outlet check valve first, followed by the spring and primary cup. Slide the piston in and secure it with the retaining ring. On master cylinders with a booster, be sure the rubber O-ring is on the piston before inserting it in the bore. Note also how all seal lips face forward— toward the radiator.

On dual master cylinders, install the secondary piston first, then the primary piston—the spring ends go in first. If you're not sure about which is which, refer to the drawings. Then, on dual cylinders, replace the springs, valves and tube seats on the primary- and secondary-system outlets. Don't forget to install the bleeder screw. To complete the assembly, install the dust boot.

I don't discuss the power-brake booster simply because it's beyond the scope of this book. If your car has booster problems, purchase a new or rebuilt unit. You'll be time, money and trouble ahead.

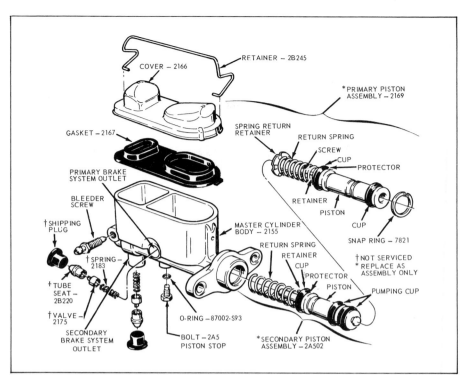

Dual master cylinder for disc brakes has large fluid reservoir at rear. Additional fluid is required for disc brakes which are at front. Dual master cylinders for four-wheel drum brakes have the same size reservoirs. Courtesy Ford Motor Company.

While honing bore use liberal amounts of brake fluid. This will help prevent hones from loading up with metal particles. As you work hone in and out, be very careful not to pull it from bore or allow it to bottom.

Parts from master-cylinder rebuild kit are from left to right: outlet check valve, spring, primary cup, piston, boot and retaining ring.

7

Engine & Engine Compartment

Before removing the engine, you should have taken a few pictures so it can be put back in original condition. Maybe it didn't look as good as this '69 G.T.500's 428 V8, but you get the idea. Photo by Ron Sessions.

Before your engine was removed, you should've photographed it so it can be returned to its original configuration. If you haven't removed it yet, photograph the engine and its compartment from several perspectives. Assuming it hasn't been *violated* through some modifications, the engine, accessory drive and surrounding area under the hood should be returned to its original factory condition. Or, if modified, components should be official Ford or Shelby offerings of that era. Also use photos or drawings for replacing stickers or decals on components or locations such as the air-cleaner housing, valve covers and spring tower. Old magazine or factory-literature photos of the cars when new are great for this. And, then, there's all the other stuff:

wiring, hoses, brake lines and so on.

After the engine is out, it should be returned to its fresh-off-the-assembly-line condition—at least in external appearance. It should also be in good running condition. If not, a rebuild is in order. Consider doing this yourself. Let's look at engine rebuilding first.

ENGINE REBUILD

The subject of engine rebuilding deserves more than half a chapter. Fortunately, HPBooks has obliged us with some *complete* books on the subject. The first is *How to Rebuild Your Small-Block Ford* and is written by one of the foremost authorities on the subject: Tom Monroe. Tom is a Registered Professional Engineer, Member

of the Society Of Automotive Engineers (SAE) and a former Ford engineer. His own 1966 High-Performance Mustang is an absolutely awesome piece of machinery, page 102. If your Mustang sports a small-block Ford V8 under the hood, you need his book as a companion to the one you're reading now.

Tom's book covers the 221, 260, 289, HP289, 302, Boss 302 and the 351W small-block Fords. It tells you how to find out what needs fixing, how to tear down an engine, repair and recondition it and finally, how to get it back into the car. It has an excellent chapter on parts interchange and identification, which is worth the price of the book. Not sure when the engine was built? Can't find the part you need? Check

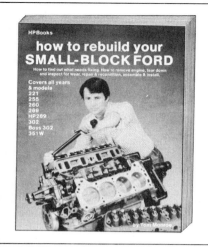
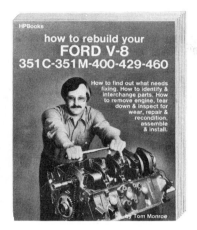

The definitive works on rebuilding Mustang V8 engines, from 260 to 429, are published by HPBooks. See inside back cover for ordering information.

OFFICIAL SHOP MANUALS

For additional information concerning your Mustang get the official Ford shop manual. Single-volume manuals covering specific vehicles were printed for cars prior to 1970. These single-volume manuals were superseded by five-volume manuals that cover *almost* all car lines.

Manuals may be available from Helm. To order, or for additional information, contact Helm, Inc., P.O. Box 07150, Detroit, Michigan 48207. Telephone: (313) 865-5000. Make sure you include year of your Mustang when ordering.

the interchange and identification section. There may be parts from a different engine that will do the job...and keep your engine original in appearance. Want to use a taller piston for additional compression? The interchange chapter will tell you what can be done and what to avoid. Don't try to rebuild your small-block Ford without this book!

If your pony happens to be powered by a 351 Cleveland, Tom Monroe has also written a book for you. Called *How to Rebuild Your Ford V8*, it covers rebuilding the 351C, 351M, 400, 429 and 460. Although most of these don't apply to the Mustang, all you're interested in is how to rebuild the engine that's in your car.

Don't feel slighted if your car is powered by a big-block. HPBooks hasn't forgotten you. *How To Rebuild Your Big-Block Ford,* by Steve Christ, is every bit as informative as the other Ford engine books. Get this one if you have a big-block. You'll be glad you did. Unfortunately, HP doesn't have anything on Ford sixes. If you have a six, you'll have to get a shop manual for your car. A swap meet or Mustang mail-order house is the best place to find early manuals.

If it's all covered in these rebuild books, why read about it in this chapter? Well, there's a lot to know about what goes on behind the scenes, what's new in the marketplace, how to detail the engine for show or for your own "styling," and a number of other bits and pieces of information worth knowing. So, let's talk about them.

How to Rebuild—There are two major approaches to rebuilding an engine: with lots of money or with very little money. Of course, there's the middle ground where most of us fall. Your budget must determine which way to go. If you're extremely fortunate, you may not have to do a rebuild. However, assuming a rebuild is in the cards, you must decide how you'll go about it. Chances are your bank account will weigh heavily on what your decision will be.

Generally, the advantage of taking the big-buck approach is longer engine life. Conversely, a low-buck rebuild usually is not as durable. How long you can expect either type of engine to last is difficult to predict.

Above, I use the word *generally* because it depends on how much work you do yourself and what shape the original parts are in. As you may know, labor is where most of your money will go. So, if you do most of the engine work yourself, checking and machining included, you can reduce the price of the rebuild considerably. This, of course, depends on your experience, available equipment and friends you have in the business. As for the parts, there's no point in boring an engine if it has little bore wear. This can also save the cost of purchasing new pistons. As for the cam and lifters, they should always be replaced if the engine has many miles on it.

Big-Budget Rebuild—Replace a part if it's questionable. The old axiom about a chain being as strong as it's weakest link applies to engines. For example, if the timing chain breaks or jumps a few cogs, the car won't go. If the old cam and lifters are reused and a lobe and lifter gets *wiped*, the resulting metal particles will ruin the new bearings in a thousand miles. This means the engine must be removed and torn down so the cam, all of the lifters *and bearings* can be replaced!

If you can afford it and you want a "new" engine, replace everything but the block, heads, crank, connecting rods, manifolds and sheet metal. In the project car, we replaced just about everything—pistons, rings, cam, lifters and bearings.

Quality Parts—Select top brands when buying new parts. This assures you the manufacturer will stand behind them. At the back of this book are listed the names and addresses of suppliers who's parts were used in the project car. All are reputable and guarantee their parts.

Know Your Machinist—Shop for the right kind of machine shop. First, the shop must be able to do the critical machine work: grinding, honing, boring and polishing. Some shops are oriented toward production work, while others specialize in custom work.

Engine shops who specialize in production work usually remanufacture *short blocks* (block, crank, rods and pistons) and

Magnafluxing is a process that uses magnetically charged iron particles to find cracks in cast-iron heads and blocks. Without this test, you might rebuild a cracked head that would be totally useless.

Bronze sleeves are inserted in valve guides. This, coupled with new valves, restores valve-stem-to-guide clearance to specifications, and is a durability improvement that helps cope with low-lead and no-lead fuel.

long blocks (cylinder heads included on short block) to sell to the industry. Walk in with your old engine and they'll give you a rebuilt one to take its place. It's quick and relatively inexpensive, but not what you should have for your restoration. Instead, your car's original engine should go back in to keep the car completely original—at least from exterior appearances. That means it should be rebuilt *for you car*. To have this done, you'll need a custom machine shop.

In a custom engine machine shop, you can talk directly to the person who'll work on your engine. You'll be able to discuss various options. For instance, there are three satisfactory ways to recondition valve guides. They can be inserted with cast-iron guides, thin-wall bronze guides or a screw-in inserts.

Spend some time selecting the machine shop you'll finally settle on. As with everything else, there are good and bad shops. The easiest way to find out who's who is to ask around. Parts houses that are not affiliated with a particular machine shop will always be able to recommend one or two shops. Even better would be a local parts house specializing in Mustang parts. Their recommendation should be considered.

Best of all, if you belong to a Mustang club, ask around. Collectively, the membership is walking encyclopedia of

Mustang knowledge. At least someone in the club should be able to recommend a place to go for engine work.

Valve Seats—Except for one necessary exception, I don't get into engine-building specifics in this book. However, because of a change in the gasoline your car must burn, you should know about an unavoidable problem.

All '65—70 Mustang engines have one thing in common; they were designed to run on "leaded" gasoline. The *tetraethyl-lead* additive in gasoline assists in lubricating and cooling the exhaust valves and seats. However, as of January 1, 1986, the tetraethyl-lead content in "leaded" gasoline became virtually non-existent under orders from the U.S. Enviromental Protection Agency (EPA). Consequently, early unhardened cast-iron exhaust-valve seats—same material as the cylinder head—are vulnerable to excess wear when modern "leaded" or unleaded gasoline is burned.

As usual, with anything government-related, it's going to cost you money to correct the problem. This is done by installing hardened exhaust-valve seats; the old seats are counterbored and new hardened inserts are installed in their place. When you ask your engine machinist about this, he'll know exactly what you're talking about…and he'll know how to do it.

Engine Assembly—Once the machining

Mustangs of '65—70 vintage were designed to run on leaded fuel. To prevent valve recession, install hard seats in heads. Photo by Ron Sessions.

is done and parts are inspected and reconditioned or replaced, you must decide who will reassemble your engine. Consider that if you follow the directions in the HPBooks' rebuild book that applies to your engine, you'll find that the job is a snap. These books carry you through each process step by step. Just pay attention, be careful and don't rush the job and all should be OK. The operative word here is careful.

No matter how careful you are, there's the chance of making a mistake, even if you're a pro. If you make a mistake, you correct it on your time and money. I've seen engines that came in and out so many times they should have been retained with zippers instead of bolts. If a machine shop assembles your engine, it should stand behind its work. A reputable shop will pay labor expenses to remove and reinstall an engine they assembled wrong. Don't forget to ask if your machine shop makes this guarantee. If they won't put it in writing, find another shop.

Regardless of who assembles your engine, it should be painted official Ford blue. Do this before you install the carburetor, exhaust manifold(s), fuel pump, and water-temperature and oil-pressure senders. The water pump and thermostat and its housing should be installed. Also install an old set of sparkplugs to keep the threads and sealing surfaces paint-free. Mask the crankshaft flywheel flange, fuel-pump and carburetor-mounting surfaces, but paint everything else just like the factory did. Don't worry about being neat; the factory didn't. If your engine has an aluminum intake manifold, like that on a Boss 302, don't paint the manifold. Mask it, too. Read on for more details on painting.

ENGINE & ENGINE-COMPARTMENT DETAILING

For many restoration enthusiasts, parts detailing is a favorite job. It's possible to take the ugliest, greasiest, dirtiest part and quickly turn it into a work of art. There are successful businesses that do nothing but remove an engine, detail it and the engine compartment, then reinstall the engine. So, if you have your engine out, and possibly rebuilt, now's the time to do this very rewarding restoration job.

There's considerably more to preparing parts than simply cleaning them in solvent and painting them. Solvents won't remove rust. And using the right paint is important. What about a primer? Some people swear by powder-coating. With these questions in mind, let's take a look at some of these things and learn how the pros do it.

Cleaning Parts—The first order of business in parts cleaning is to remove the dirt and grease. Once you've scraped off the caked-on deposits, an automotive-type cleaning solvent is best for finishing the cleaning operation.

Use a solvent that leaves no oily film. This removes kerosene from the list right off. Most large auto-parts stores sell parts-cleaning solvent. A satisfactory replacement for this solvent, if not available, is

The more parts you replace with new, the longer your engine rebuild will last. By all means, replace the timing chain and sprockets for long wear. I use the Cloyes True Roller chain. Photo by Tom Monroe.

With engine out of car, you can do a thorough job of cleaning, painting and detailing engine compartment. Photo by Tom Monroe

paint thinner. This is always available at commercial paint stores where they cater to the house painter. Buy the lowest grade possible; the kind the painters would use to clean their brushes. Don't use lacquer thinner, it's too expensive and too combustible. Likewise, don't use an auto-enamel thinner. This is very, very expensive, often three to five times as much as an enamel thinner for house paint.

Solvent Tank—You can build a solvent tank for very little cash outlay. Start with a good, leak-proof, 55-gallon steel drum. Then make a parts basket that will fit inside and hang below the lip of the drum about 12 in. For the basket, I use the bottom 6 in. of a 30-gallon drum, punch holes in it and weld three hooks to the sides. The basket then hooks to and sits inside the 55-gallon drum.

Remove the top of the 55-gallon drum, if it has one. Stand the drum upright and fill it with water to within 24 in. of the top. Fill

Beadblasting is easier on cast-aluminum parts than sandblasting and leaves a satin finish.

OK. So maybe you're not going to pull a perfectly good running engine just to detail the engine compartment. First, warm the engine to operating temperature and spray on commercially available degreaser. Note carburetor wrapped in plastic bag.

Small parts are easily stripped of paint and rust with rotary brush attachment of bench grinder.

the remainder of the drum with solvent. The solvent will float on top of the water. Now, put your greasy parts into the basket and immerse the whole works in the solvent. The grease and dirt that come off will pass through the solvent and settle into the water below, keeping the solvent clean and fresh. Cover the drum when not in use and the solvent will last for a long time.

Chemical Dipping—For large parts (car doors up to car bodies), chemical dipping is the most effective for removing grease, paint and rust. This is like furniture stripping, only the chemicals are stronger and the tanks larger. You bring in the piece to be stripped. A business that specializes in this is Redi-Strip. They have two processes: one for removing paint and similar coatings such as tar, plastic body filler and rubber, and another for removing oxidation—rust. If you can't find a Redi-Strip facility, write Redi-Strip Paint Stripping & Rust Removal, 9910 Jorden Circle, Santa Fe Springs, CA 90670.

The part to be stripped is submerged in a vat where all but the metal is removed. The chemicals are then neutralized, the part dried and finally primed. This is great for body and engine sheet-metal parts. Unfortunately, not every community has such a service. If it's available, use it. Otherwise, you may have to use sandblasting.

Sheet-metal engine parts, all of which are smaller than most sheet-metal body parts, can be stripped by your engine machinist. He'll dip the parts in his *hot tank* which is filled with a caustic solution, or he'll put them in his jet-spray cleaner where the parts are sprayed with a similar solution. Grease, dirt, paint and even phosphate and cadmium plating is removed, leaving nothing but bare metal. So, if a part

is plated, don't use this method. Ask first. Re-plating is much more difficult and expensive to have done than removing it.

Sand/Beadblasting—Another way to get parts clean is by sandblasting or beadblasting. In sandblasting, sand is sprayed with compressed air against the part. This quickly removes *everything*, including some of the metal. If you decide to sandblast a part, be sure all the grease is removed first. Otherwise, grease will be forced into the pores of the metal. When paint is applied, the remaining grease will create *fisheyes* in the paint, small circles where the paint did not flow smoothly. So, sandblast only as a last resort. Use beadblasting instead.

Like sandblasting, beadblasting is done with compressed air. Instead of sand, however, small glass beads are used. Their sharp edges cut through paint and rust, but don't affect the metal—unless it's soft aluminum or pot metal. And even these softer metals can be successfully beadblasted as long as a continuous stream of beads is not concentrated in one area for too long a time. Therefore, beadblasting is my choice for removing paint and rust.

Beadblasting is done inside a cabinet with the operator working outside. The hose and part to be beadblasted is held by the operator using rubber gloves and sleeves that extend into the tank. This keeps the expensive glass beads contained

Then, steam-clean engine compartment. Many gas stations perform this service. Or use high-pressure spray at coin-operated car wash.

Waterless hand cleaner does an amazing job of cleaning sparkplug wires, wiring harnesses, hoses and anything plastic or rubber.

so they don't end up in the air you breathe. Glass beads, shards and fibers are very damaging to the lungs. *Warning: Do not use glass beads in your home sandblaster. It could cause silicosis of the lungs or other life-threatening illnesses.*

Just as you would when painting, mask areas of a part that you don't want sandblasted or beadblasted. Such areas include threads or irreplaceable tags or decals.

Finishes—When you've cleaned all of the parts, determine which is the correct finish for each. Generally, the finish will be paint. There are a number of other finishes available that can or should be used in the restoration of your car. These include powder-coating, sometimes called *powder painting,* phosphating, and cadmium, zinc and chrome-plating. Most of these have a place in the restoration of your car.

On page 171, I've indicated the correct finish for each part. If you are restoring your car to original condition for competition in car shows, the finishes must be correct. Many people, however, like to restore to a little better than original factory condition. To that end I'll discuss the above finishing processes. Turn page 119, for a list of the original factory exterior color codes.

Phosphating—What engine and engine sheet-metal parts weren't originally painted blue, orange or gold were painted

Cover important decals with masking tape before painting. Engine compartment and inner fenders go semi-gloss black.

Ford chassis black or were phosphate-coated. Phosphated parts, such as hood hinges, sometimes rust. To return such parts to original factory specs, beadblast them, then have the parts phosphate-coated or, if you can't find someone to do the phosphating, paint them with clear acrylic.

Look in the Yellow Pages book under PLATING to find a business that does

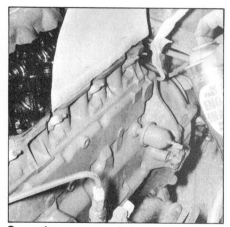

On engine, remove what you can to improve access to components and protect from overspray. Use high-temperature enamel for engine as per the chart, page 171. Don't paint oil-pressure sender as was done here!

phosphating. Expect to pay a premium because such businesses deal mainly with commercial accounts. I hate to say this, but your little batch of prized Mustang parts represents more trouble than profit to them.

Paint—The longest-wearing paint is a polyurethane with a catalytic hardener. It should be applied over a zinc-based primer to prevent rust. Although this is a bit of

101

Not bad for an in-car job, eh? Six is much easier to detail in-car than a V8, particularly a big block.

Small parts are best painted off the vehicle.

overkill, it makes a superb job. If you don't like the high-gloss effect, add some flattening agent to the paint. This will keep the ultra-smooth finish that is so easy to clean, but will reduce light reflection for that original semi-gloss appearance.

A less-expensive paint is synthetic enamel with an added hardener. Eliminate the separate hardener and you have a very satisfactory and inexpensive paint.

For convenience rather than longevity, use an aerosol can. This does an excellent job, is used in most restoration shops, dries in minutes, and lasts fairly well, although not as well as the paints just described.

Powder-Coating—If you're fanatical about doing the most durable job, powder-coat parts where possible. Powder-coating lasts longer than polyurethanes, is virtually impenetrable by rust, and nearly lasts "forever." Unfortunately, it's not a home process. The part must be sent out just as if it were being plated.

To prepare parts for powder-coating, remove any dirt and grease, then send them to the powder-coater. He'll strip them, then coat them. The *electrostatic* power-coating process is as follows:

Clean parts are suspended from a hook that's negatively grounded. Positively charged powder particles are applied with compressed air. This electrostatic charge "locks" the powder particles to the parts. They are then placed in an oven and baked

After engine compartment is detailed, it's ready to receive rebuilt engine. Notice that decal on driver's side spring tower is not installed horizontally as it should be, but is the way it came from the factory. Reinforcement at center of cowl is used with engine-compartment export brace. Photo by Tom Monroe.

at a temperature of 350F. This melts the powder, fusing it to the metal. The resulting finish is extremely durable. Any metal can be powder-coated as long as its melting point is above 350F. This, of course, eliminates anything that includes rubber, plastic, lead, solder or any other low-temperature alloy. If powder-coating appeals to you, visit a shop and take a look at a finished sample. It's available in gloss, semi-gloss and flat.

Electroplating—Many restorers like the

Detailing has become a science, as demonstrated by Dave Finner's prize-winning '68 fastback. Note phosphate-finish fender bolt heads, "AC OK" stamp for air-conditioning test at the assembly plant, and original-type Autolite battery *with* tag. Photo by Ron Sessions.

Part of the fun of detailing is to restore all of the original decals and stickers, such as this Autolite coil label. Reproduction pressure-sensitive labels are available from Jim Osborn Reproductions, Lawrenceville, GA.

convenience of electroplated finishes. They are long-lasting, easy to clean and, in the case of all but chrome, are quite inexpensive. As with phosphating, look for a plater under PLATING in the Yellow Pages. Following are common types of plating.

Cadmium-plating, commonly called *cad-plating,* was once the best choice for all metal parts. With the use of more and more aluminum and plastic parts, cad plating fell out of favor with the manufacturers. About the only cad-plated products on Mustangs are the fender bolts along the inner fender under the lip of the hood.

If you choose to use cad-plating, it's very inexpensive. Usually, the plater will charge you by the pound for small parts such as nuts and bolts, screws, brackets and other minor pieces. Large items will be charged for by the piece.

You can take your dirty and greasy parts to the cad-plater. They're chemically cleaned before being plated. When they come back they will be bright silver, gold or whatever you specify.

Zinc-plating is an alternative to cad-plating. It's getting more and more difficult to find someone to do cad-plating. Fortunately, for about the same price, you can have your parts zinc-plated. Zinc is more susceptible to corrosion, but nevertheless may be a solution to plating some of your Mustang parts. But if you have the choice, go with cad-plating if it's available.

Chrome-plating is the queen of electroplating processes. As such, it's also the most expensive. The cost is mostly in the labor required to polish the metal prior to applying the nickel and chrome. If the chrome shop does a good job, your part should be virtually rustproof.

Very little was chromed on Mustang engines. Mostly, it was reserved for high-performance-engine dress-up kits. Parts included the valve covers and air cleaners. Chances are the original chrome will be OK, so rechroming shouldn't be required.

These are your options as you detail the engine and engine compartment. If you plan to show the car or wish to sell it for the highest price possible, use the original finishes when doing your detailing. See the chart, page 171. However, if you want to raise your car out of the ordinary and don't give a hoot about anything but satisfying yourself, try some of the more exotic detailing tips just described.

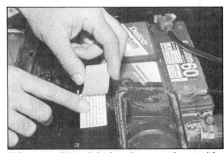

When applying labels, clean surface with alcohol and align them carefully. Adhesive is powerful and once label is down, it won't come off without self-destructing. Never mind GM battery and aftermarket radiator hose and screw clamp—this Mustang's in transition from daily driver to restoration.

Eastwood stainless-steel coating returns cast-iron exhaust manifolds to original, unpainted appearance. Coating will survive where high-temperature paints won't. Photo by Ron Sessions.

8
Body & Paint

Body and paintwork is perhaps the most exciting part of the restoration process. Armed with a few body tools, paint and body-prep supplies, a welder and a spray outfit, your Mustang's flanks will soon be as good as new. Photo by Michael Lutfy, courtesy of *Mustang Illustrated* magazine.

The final appearance of the paint tells the world what kind of restorer you are. A flawless job bespeaks the professional; a haphazard job, the amateur. There is no reason why your paint job cannot be flawless and appear professional. The difference will be in the time it takes—many times that of the professional if this is one of your first paint jobs. Work step-by-step until you're completely satisfied with your Mustang's finish. You'll be rewarded with comments such as, "Where did you get that great paint job?" Of course, even the best paint cannot cover up substandard bodywork. All panels must be straight, free of excessive filler and structurally sound.

As with engine overhauls, paint and bodywork cannot be covered in one chapter. As mentioned earlier, HPBooks' *Paint & Body Handbook*, written by craftsman Larry Hofer and your's truly, covers all aspects of body repair from minor dents to major accident repairs. In the paint section, we carry you through all steps necessary to achieve a flawless finish. If you're truly serious about the best paint job you can do, this book's for you.

BODYWORK

In this section, I'll walk you through a complete body restoration. You may not need to be this thorough, but it will bring the car as close to original as is possible.

If you have not completely disassembled your car, now is the time. This will let you inspect every nook and cranny for rust, collision damage and previous repairs. More car bodies die of rust and corrosion than any other cause. Iron ore to iron oxide—like ashes to ashes and dust to dust, all it takes to reduce a car body to a pile of rusting iron and steel is a little bare metal, time, water and oxygen.

Many body parts are still available from Ford or are being reproduced by an aftermarket company. And there's nothing wrong with using sound, unrusted used panels from your local salvage yard. The southwestern portion of the U.S. is filled with them. Do beware, though, of sheetmetal prices that seem to good to be true.

Check every nook and cranny for rust. On double-panel sheet metal, such as front fenders, inspect lower portion of inner panel for perforation, as shown here.

Luckily, Mustang was a mass-produced car and most sheet metal is available either from Ford or the aftermarket. Make sure, however, that part is from a quality producer, as was the case with this inner fender.

Many of my fellow hobbyists have complained about the inferior quality, fit and finish of replacement panels imported from across the Pacific. Do yourself a favor and stick with original-quality sheet metal for your Mustang's restoration. There's nothing worse than saving a few bucks on a cheap panel, only to spend countless hours with a hammer, dolly, drill motor and wood block trying to make it fit. For my money, go for the panel with the *"Made in the USA"* label.

PAINT REMOVAL

Before you can really see the small rust areas, you should remove the old paint. This is a must if there is more than one coat of paint on the car. You cannot build a good foundation for new paint over *more than one* original coat. Likewise, a second coat can hide body damage as well as rust. Got a 3-in. dent in that fender? Slap in a little plastic body filler, file it down and paint it. Who's to know? You'll know when it falls out, carrying your expensive paint job with it! And if you are going with a non-stock lacquer or other type of custom paint, it may lift previously applied coats of enamel. So really, bare metal is your best friend.

Yes, stripping all that paint off is a real chore, but it's your only assurance that the job will be right and will last. There are exceptions, of course: If you're looking at factory-original acrylic enamel that may be a little thin, but otherwise is free of cracks, crazing, blisters and other defects, wet-

For stripping paint from small areas, Eastwood's rotary cleaning disc does job with no muss or fuss. Photo courtesy The Eastwood Company.

Removing old paint with stripper is serious business. Be prepared. Photo by Michael Lutfy, courtesy of *Mustang Illustrated* magazine.

sand, prep-sol and paint over it. The factory-baked sealer coat beneath that original paint is likely better than anything you'll ever apply. And remember, anytime you expose large areas of bare metal to the atmosphere, you're giving rust a chance to gain a foothold.

Body Dip—Nevertheless, there are a number of ways to remove paint from metal. The best way is to dip the body in chemical solution. It is, however, the most expensive. If a facility is available in your area that can dip the entire body in paint-and-rust remover, that's the way to go.

The entire car must be disassembled. The only thing that can be dipped is the body. Everything else must be removed— wiring, rubber, plastic, glass...everything.

The easiest way to transport a disassembled body is to leave the rear end, wheels and tires on the car. Rent a trailer designed to haul a car by the front wheels only, such as used behind big recreational vehicles. The front of the car is supported by the trailer, with the rest of the car supported on its rear wheels.

Load the fenders, doors, valances, quarter-panel extensions, headlight doors,

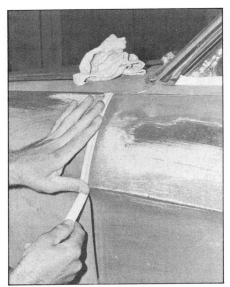

Before applying stripper, tape over gaps between panels to keep stripper from attacking weatherstripping and paint in door jams. Photo by Michael Lutfy, courtesy of *Mustang Illustrated* magazine.

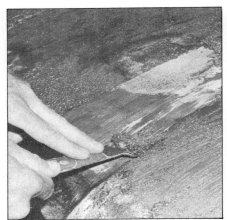

Strip paint outdoors, if possible, to ensure proper ventilation. Use heavy coats and scrape it off before it dries. C'mon, don't forget those rubber gloves. Photo by Michael Lutfy, courtesy of *Mustang Illustrated* magazine.

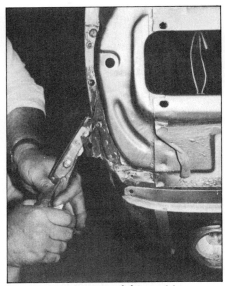

In some cases, you might want to replace a section rather than replace an entire panel, such as on a door skin or rear quarter panel. Eastwood's Mini Nibbler cuts through sheet metal without distortion. Photo courtesy The Eastwood Company.

headlight buckets and all other pieces you wish to have cleaned into the car and tow it down to the dipper. There, the operator will support the body on a cradle and hoist it off the trailer, allowing you to disconnect the rear-axle assembly from the body and load it back on the trailer. From there, the dipper folks will take care of your car body and the loose parts. In a week to ten days when you get the car back, all of the paint, body sealer, grease, dirt and rust will be gone and the body will have a fresh coat of primer. You can then begin your bodywork.

Brush-On Stripper—The second method is to remove the paint yourself using a brush-on paint stripper, such as Klean-Strip Aircraft Remover. This Ditzler product, code DX-586, is the most effective I've found. It can be purchased at most automotive paint-supply houses that carry Ditzler paints. Get a pair of plastic gloves to wear when you apply this stuff—*it burns the skin upon contact!* And the acid-like fumes given off by the paint stripper can damage your lungs if inhaled in high concentrations, so make sure your work area has excellent ventilation. Do the job *outside* if weather permits.

Apply Aircraft Remover with a brush, wait a few minutes until the paint lifts, then scrape the paint off with a razor-blade scraper. This tool is available at paint-supply houses. You can also use a hard-plastic scraper used for removing ice from windshields in Northern climes or a wide plastic body-filler spreader. These will not gouge the metal as will a metal scraper. But on a large car, you may need more than one ice scraper or spreader, as they dull quickly. The stripper is water-soluble and can be washed off the metal surfaces when all paint is gone.

One of two steps should be taken immediately—unless you live in the desert and it's the dry season. If you have to leave the bare-metal surfaces outside, especially overnight, they must be sealed from rust. The simplest way to do this is to apply a light coat of fresh engine oil or ATF—which, of course, will have to be completely removed before doing any body or paintwork. This will prevent the metal from rusting in all but a rain shower. If you are working indoors in a dry climate, the metal may be left bare for one or two days. Longer than this and the metal will begin to rust from the humidity in the air.

If extensive metalwork must be done on a body panel, leave it bare metal. Bare metal seems to be easier to work on than primed or painted metal, especially if you plan to do any leadwork. If no bodywork is expected on a part, it should be prepped and primed.

PREPPING

Prepping a panel simply means using an acid etch to clean it and give a "tooth" to which the primer can cling. Use one of the available metal preps, such as DuPont or Ditzler. Mix it with water according to directions. Wearing rubber gloves and, using very coarse steel wool, apply metal prep to the bare metal and scrub with steel wool. When all remaining paint, stains and rust are gone, wipe off the metal prep with a clean, dry rag or towel.

The directions don't suggest it, but I like to then wipe everything down with a generous wash of lacquer thinner. This removes any acid you might have missed with the dry towel. After this final operation and when you're sure all surfaces are thoroughly dry, apply a coat of primer. You can do bodywork over primer just fine. If you choose to do leadwork, you'll have to grind the primer away to bright metal for the lead to adhere.

RUST REMOVAL

On a 20-year-old car, such as an early Mustang, rust can be a big problem. For a convertible, it's an even bigger problem. Like it or not, most convertible tops leak. If the carpet isn't pulled up and dried, rust can get a foothold in the floorpan. If left unchecked, rust can spread to the rear-spring perch and shackle mounts. And remember, the floor is part of the structure on a unit-

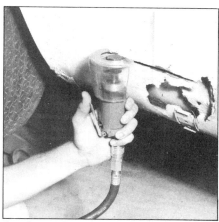

Another option is a pneumatic cut-off tool. Highly maneuverable, it slices easily through sheet metal. Photo courtesy The Eastwood Company.

On double-panel parts such as doors, you can grind away spot-weld flange, separate panels as shown, and cut out rusted section.

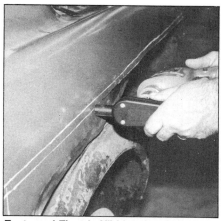

Eastwood Electric Nibbler zips through a Mustang quarter panel. Drill-powered nibbler does job with reciprocating die. Just drill 1/2-in. hole, insert nibbler head and go. Photo courtesy The Eastwood Company.

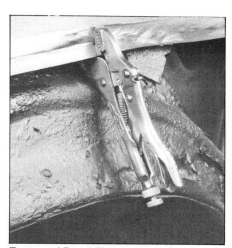

Eastwood Panel Flanger Tool offsets metal into a coach joint, allowing flush welds. Photo courtesy The Eastwood Company.

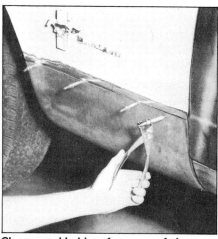

Cleco panel holders free area of clamps for welding. Drill 1/8-in. hole through both panels and install holders. Photo courtesy The Eastwood Company.

Replacement panel can then be gas, MIG, arc or spot welded in place. Shown is popularly priced Eastwood Panel Welder that operates on 110-volt house current. Photo courtesy The Eastwood Company.

body car, and the integrity of suspension locating members is a definite safety item. Other places to look for rust on any Mustang bodystyle—notchback, fastback or convertible—are along the rocker panels and torque boxes, lower edge of the doors, lower edges of the wheelwells and quarter panels and the trunk floor. In severe cases, the rear springs can come crashing through the trunk floor due to trunk-floor and rear-"frame-rail" rust, gas tanks can fall out due to trunk-floor rust, and convertibles can fold in half due to rocker-panel and floor-

pan rust. So beware.

Oddly enough, most rust is on the right side of the car. This is a result of the crown on many streets and highways. Especially when parked, water will run to the passenger side of the car because of this crown. It collects there and rusts the metal. Although cars that spent most of their days parallel parked on streets in areas of the country that use much road salt in the winter tend to suffer accelerated corrosion on the left side due to salty slush being deposited there from passing traffic. So don't overlook any

areas of a Mustang when searching for rust.

There is only one correct way to treat rust; cut it out and replace the metal. Yes, many magazine article show you how to use fiberglass to patch the hole. You probably know one or two bodymen who swear by the "T-shirt and Bondo" method—wad-

Back to the door example, you can make your own panel on flat sections with sheet steel. Cut new panel to fit section.

To prevent warpage, weld seam is hammer-welded. This keeps panel flat and and brings weld seam level with metal. Job is finished by edge-welding skin to door frame, filling and priming area, then painting.

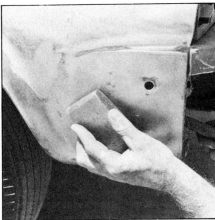

Small dents in accessible areas respond to metalworking with hammer and dolly. Photo by Michael Lutfy, courtesy of *Mustang Illustrated* magazine.

up an old T-shirt, stuff it in the rust hole and cover it over with Bondo. This works fine for a few months. Then the Bondo begins to crack out while the T-shirt soaks up more water, and the rust hole gets bigger.

There's really only one way to go; replace the metal. That means either bolting or welding in a new piece or patching in a repair section. Obviously, if the damage is severe and the piece is something that bolts into place—fender, hood, trunklid, door and so on—why not replace it, especially if the piece is still available, reasonably priced, and you're restoring this car to keep. On the other hand, if damage is minor and easy to gain access to, or if the part is no longer available, prohibitively expensive or an integral part of the car, weld in a repair section.

RUST REPAIR

Before you can cut back to good, clean metal, you must be able to get your tools into the area. On double-panel pieces such as doors, hoods and trunklids, one option is to "reskin" the panel—install a new outer panel over the inner panel.

Take a door, for instance. I use a body sander with a #36-grit disc to grind away the radius of the crimp—*hem flange*—holding the door skin to the door frame. Or, on spot-welded panels, you can use a special drill bit, such as that available from the Eastwood Company, that drills out the spot welds holding the two panels together. After using either method, you can then lift the skin up to work on it. The next step is to cut away the rusted area.

Because the door rusted from the inside out, check the back of the door skin for surface rust. Cut far enough into clean metal to be sure you get all of the rust. A pair of straight-cutting aviation shears does a fine job. Now, cut a new piece of metal to fit the hole.

Lay a piece non-galvanized sheet metal, sometimes called *black metal*, under the cutaway area and scribe out the desired shape. Cut and trim this to fit. If you were working a curved area here, you would want to shape the new metal to the curve before fitting it.

When the new piece of metal fits the cutout, you're ready to weld it in. Be sure the fit is snug. Try to avoid gaps of more than 1/32 in. This will make the welding easier.

Using an oxyacetylene welder, tack-weld the new piece in place. Use the smallest tip you have to keep heat low and warpage to a minimum. A 00 or 000 tip works well. If you get any warpage, straighten it before you proceed. When everything is well-aligned, weld the new piece in completely. Distortion will be greatly reduced if you weld the new piece in with a MIG—wire-feed—welder or one of the newer popularly priced spot welders. For details on welders and welding techniques, see HPBooks' *Welder's Handbook*.

After welding in a section of new metal, the panel probably looks like a stormy night on the North Atlantic! This is the result of buckling from the heat. The warpage can be straightened out by *hammer welding*—heating the affected area cherry red and

Plastic body filler is perfectly acceptable for filling small body imperfections. It's inexpensive, easy to use and, if correctly applied, durable. As with leadwork, apply plastic filler only to clean metal, free of all rust and paint. Photo by Michael Lutfy, courtesy of *Mustang Illustrated* magazine.

working it with a hammer and dolly. To do this, it's best to commission the help of a friend.

Begin by having your friend heat an area on the edge of the weld. This should be about 3/4 to 1-in. in diameter and a deep red-orange. The area will rise, forming a cone. With a hammer and spoon or dolly, *quickly* flatten this cone and the surrounding area, being sure that the weld bead has been flattened in the process. It should be flush with the sheet metal. Proceed along the weld bead in this manner until the bead and surrounding area are flat.

The rest of the operation becomes fairly obvious. Trim the excess metal from the corner, clamp the door skin and door frame together and weld along the seam. Again, use as little heat as possible; the result of the finest tip that will do the job. Finish by grinding the weld until it smoothes out a bit.

You've now done the correct job of repairing a rust area anywhere on your Mustang's body. If the area is very large, it may require drilling out spot welds to lift the sheet metal. You will find this situation on a rear quarter panel just behind the rear-door pillar. The fender skin can be accessed by drilling out the spot welds along the door pillar and under the quarter panel where the skin is welded to the inner fender. Be sure to leave enough material on the patch to weld it back to these weld flanges.

Disc grinder removes filler fast. Be careful not to remove too much. Photo by Michael Lutfy, courtesy of *Mustang Illustrated* magazine.

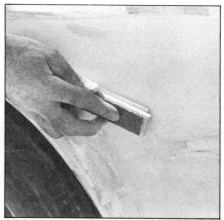

To find imperfections, glue 80-grit sandpaper to flat stick . Photo by Michael Lutfy, courtesy of *Mustang Illustrated* magazine.

REAR FRAME RAIL & TRUNK FLOOR REPLACEMENT

Rusted trunk floor and rear "frame rail" of '69 Mach I, page 32, comes into Tucson Frame Service for major reconstructive surgery. Photo by Ron Sessions.

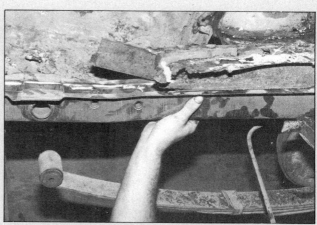

After removing rear bumper and valance, draining and lifting out fuel tank, disconnecting rear suspension and removing rear wiring harness, Doug Schoneck trial-fits replacement "frame rail".

Further damage—stress cracks ahead of rustout section will have to be welded.

As a reference, measure from center of axle assembly to spring-eye bushing on "good" side. Photo by Ron Sessions.

Cut is made with die grinder with cut-off wheel.

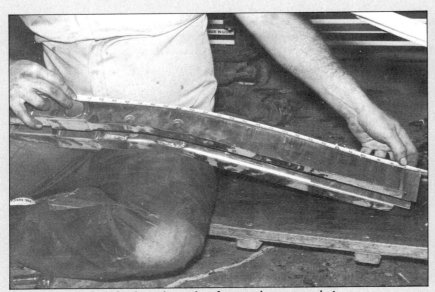

Then, measure length of repair section from spring-eye socket.

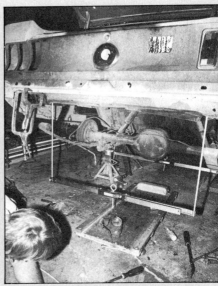

New frame section is clamped in place and three-point frame alignment fixture hung. Rearmost fixture hangs from rear leaf-spring eyes. Photo by Ron Sessions.

Hardware from rusted section (left) is transferred to repair section.

With all alignment points set, rear frame rail is MIG-welded in position. Wire-feed welder works well in tight quarters. Photo by Ron Sessions.

Drill trunk-floor section for fuel-tank mounting screws and puddle-weld holes to simulate spot welds.

To weld in trunk-floor section, fuel tank was jigged in place as an alignment fixture. *Tank was drained, flushed out with alcohol and purged of all vapors prior to welding.* Never weld near a fuel tank that contains gasoline vapors—the chance of an explosion is high.

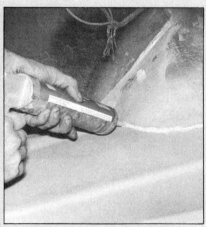

As with all other panels, caulk or recaulk seams to keep moisture out of joints.

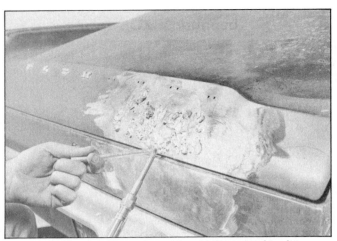

As a body filler, lead is preferred by old-timers and purists. After cleaning all rust and paint from repair area and applying tinning compound, small blobs of lead are melted from bars onto surface with torch.

Beeswax-coated paddle is used to smooth blobs of reheated lead.

LEAD WORK

As long as you're doing excellent repair work, why not finish it off with lead rather than a plastic filler? There are only two secrets to good lead work: clean surface and correct lead consistency. Get these two right and folks will think you're a 70-year-old lead worker.

There are two schools of thought on the use of lead filler. As stated previously, one is that lead is best for all applications—more durable. The other conviction is that one should use the same filler on a restoration as was used originally by the factory—lead filler on early Mustangs and plastic—yes, plastic—on later-model Mustangs. You be the judge.

Surface Preparation—To do leadwork, the surface of the metal must be hospital clean to get a good bond, even with a good flux. Grind the metal down with a #80 disc on a high-speed grinder, a 3-in. drill grinder or standard body grinder. Remove all stains, dirt and weld slag. Use a small triangular file to get into weld-area nooks and crannies. Clean the area out to about 4 or 5 in. beyond the weld. This area must be spotlessly clean. Any dirt or stain will refuse to bond to the lead. This will result in cracks. Moisture will enter here and it's rust time all over again. Get it clean. The next step is to tin the area.

If you don't have them, you'll need to pick up a few tools and supplies: tinning

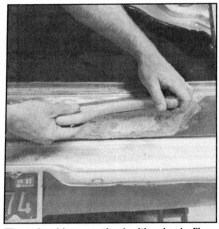

Then, lead is smoothed with a body file.

compound, a lead paddle, beeswax, two or three sticks of lead and two body files—one straight and one curved quarter-round. Begin the tinning process by warming up the area to be tinned.

Apply Tinning Compound—Pass the torch over and around the area to be tinned, slowly raising the temperature to a point where lead will melt when it is applied to the surface. While wearing a leather glove, scoop up a glob of tinning compound on a ball of very coarse steel wool and apply this to the heated area. Without gloves, you'll burn your fingers. The tinning compound will melt and flow over the surface as you pass the steel wool around. Repeat this procedure until the entire area has a thin coat of tinning compound over the entire area. It should be as shiny as a pool of mercury. Now you're ready to apply the lead.

Leading—Again, this is a bit of a trick. Reheat a small area until the tinning begins to melt and the bar of led melts when touched to the metal. If you get it too hot, the lead will run off; too cold, and it will not bond properly. Heating the metal and the stick of lead at the same time, make little mounds of lead about 1-in. apart all over the surface to be covered.

Put down the stick of lead, warm the lead paddle and pass it over the beeswax to lubricate it. Rewarm the lead in a 3- to-4-sq-in. area, and push it around with the paddle until the lead is leveled out and there are no valleys or voids. Allow things to cool down before proceeding.

Using a body file, level the lead to the same height as the rest of the metal. Clean the area with metal prep as described previously. This will neutralize the acid flux which, if not removed immediately, will cause the clean metal to rust. Wash with lacquer thinner and prime the area.

Bare metal should be etched with an acid solution such as Metal Prep to give paint a "tooth" to adhere to. Apply with steel wool or a Scotchbrite pad and wear plastic gloves for protection. Photo by Michael Lutfy, courtesy of *Mustang Illustrated* magazine.

Body fillers come in all shapes and sizes. Each has its place. Photo by Michael Lutfy, courtesy *Mustang Illustrated* magazine.

Remember the two warnings: clean surface, and correct melting temperature of the lead. The rest is easy.

You no doubt have heard the old horror stories of babies eating chips of paint off walls in old houses and then dying. These deaths were the result of lead poisoning, caused by ingesting lead-based paint used up until about 40 years ago. This same lead is in your hands now as you go about filling voids with it. Take a few precautions here to prevent lead poisoning to yourself. Wear gloves. This will keep the lead out of any cuts. Work in a well-ventilated area to keep down the fumes. Keep your face away from the fumes as you work.

PAINTWORK

When the bodywork is completed to your satisfaction, all rust has been removed and there are no more dents, it's time to turn your attention to finishing the metal and painting it.

PAINT SYSTEMS

Expect some "sticker shock" when you go to the automotive paint-supply store and price the paint you plan to use. One of the reasons paint is so expensive is the cost of research. Every paint manufacturer hires well-educated and well-paid chemists. These chemists develop the paints you will eventually use on your car.

Chemists and engineers get credit for the shine of the paint, the way it flows over metal and its ability to stand up to the worst weather, even to "acid rain." I mention this to make a point: You're paying for this knowledge; use it. Don't play chemist yourself.

When chemists and engineers develop a paint, they also come up with a reducer, a primer, a surfacer and a sealer to go with it—a *paint system*. A manufacturer will guarantee its paint system, but may not stand behind individual parts of the system if they are used in conjunction with other manufacturer's materials. For the best job, select a manufacturer and stay with that system. It will work better for you, and the people at the paint-supply house will be able to help you with most problems that may arise.

PERFECT PANEL

Before paint is applied to the car, every panel must be shaped exactly as it originally came out of its die—flat where it should be flat and contoured where the contours should be. A paint job will only look as good and last as long as its foundation. This foundation is the perfect panel.

Over the years, minor parking-lot dings, bumps and bangs to the sheet metal cause what are called in the trade, *ripples*. They are called ripples for their similarity in appearance to ripples on the water. There is

only one way to remove them, short of replacing all affected panels: they must be filled.

Fillers are applied to the metal, then block-sanded to remove all the material above the highest part of the metal. This fills the valleys. Let's look at how these fillers are used.

FILLERS

First off, *primer* is not a satisfactory filler. Eighty percent of primer is the carrier, which consists of thinner and a binder. Only 20% of primer is solids. This does not provide much filling capacity. Primer/surfacer is a much more satisfactory filler. The ratio here is 60%/40% solids-to-carrier. This gives a much greater filling capacity than primer alone.

To reach the 80/20 ratio of solids-to-carrier, use *glazing putty*. This is applied with a spreader rather than a spray gun. For really deep fills, spot putty at a 90/10 ratio is the ticket. This is used to fill voids up to 1/16-in. deep.

Polyester surfacer and polyester glazing putty are becoming popular fillers with painters. Lacquer-based fillers described previously have one shortcoming: they dry by evaporation and consequently shrink. You apply a coat of surfacer, sand it down after an hour or so, then apply paint. In a week, you can see sanding scratches under the surfacer because it continued to shrink. Polyester fillers don't do this.

Polyester fillers harden by chemical curing. A hardener, or catalyst, is added before use. It hardens by heat and chemical reaction, much like Bondo and other plastic body fillers. This eliminates the shrinking associated with lacquer-based fillers.

Although expensive, polyester fillers such as Featherfill and polyester galzing putty such as Amac are great time-savers.

Using Fillers—Let's return to our door from which we removed the rust section to see how fillers are used. Because we did a good job of leveling the repaired area with lead, a few coats of surfacer should give enough fill to produce a good flat surface. Reduce the surfacer by 200% with an inexpensive, dual-purpose lacquer thinner, or primer/wash.

Apply two medium coats at about 50 psi at the gun head, allowing 10 minutes between coats. Let it dry for at least an hour. Now, for a little trick: Using an aerosol spray can of black lacquer, ever so lightly dust a coat of paint over the surfacer. This *guide coat* should just speckle the surface, not color it.

With a 16-in. block sander and 80-grit sandpaper, begin going over the surface in

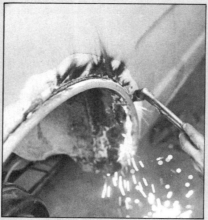

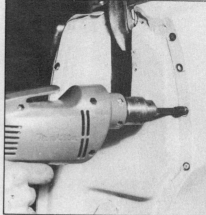

Rear quarters on Dave Cawthorne's convertible must be replaced as the wheel wells have been radiused and rust has set in. Larry Hofer, one of Southern California's best bodymen will replace the wheel well and quarter panel. Here, Larry has removed some of the plastic body filler and punched into the rust.

Often, it's necessary to cut away sections of panel to gain access to inner areas.

Factory uses spot welds to join panels. Spot welds are detected by dimpled flanges. Here, Eastwood's Spot Weld Cutter drills through outer layer of metal only, separating panels. Pilot is adjustable to different sheet-metal thicknesses. Photo courtesy The Eastwood Company.

QUARTER-PANEL REPLACEMENT

There will come a time when trying to straighten a crumpled or reconstruct a perforated rear quarter panel will be more trouble than it's worth. In this case, you should replace the whole panel rather than fuss with it. Although it may look like a rather large project, it breaks down into small segments that are easy to handle.

Removal—The basic concept of removing the rear quarter panel is to drill out all of the spot welds and remove the panel from the car. It is a bit more complicated than that, however. In the accompanying photos, you will see that Larry has used an oxyacetylene torch to cut away a section of a panel that blocked access to the spot welds. This is perfectly acceptable, providing a replacement panel is available, and makes the job go easier and faster.

Begin by drilling out the spot welds along the inside of the door frame. If you lightly sand over the tops of the welds, they tend to show up a little better. The spot weld is lower than the rest of the metal. When you sand over the top, the lower area of the weld becomes visible.

When drilling out a spot weld, you must only drill through the outer layer of metal. Drilling through the inner panel weakens it and eliminates an area to weld the new panel to. As mentioned previously, the Eastwood Company makes a special drill bit designed just for drilling out spot welds. It features a centering pilot to keep the drill centered in the weld and an arbor that can be adjusted to cut only through the outer

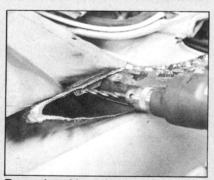

To reach welds under quarter panel, Larry cut access holes.

Further portions of panel are cut away to reach buried spot welds. See how easy it is now to reach spot welds that attached quarter panel to rocker panel.

layer of sheet metal. If you're using a regular drill, usually a 3/8-in. bit is large enough to cover the weld. Don't forget to center-punch the spot weld before you drill.

After completing the inside edge of the door, drill out the spot welds along the inside radius of the panel to separate it from the inner fender well. On the convertible, move up to the area around the top well and drill these welds. On the fastback and notchback, note where the panel joins the roofline in the enclosed line drawings and drill those welds. Unless your car had a vinyl top, this seam—called the *coach*

joint—will be lead-filled. Use your torch to remove the lead—it will liquify and pour down onto the quarter panel. Then you can access the spot welds.

Note the areas where Larry has cut away the panel to access the welds. This includes the welds where the panel joins the trunk opening and at the rocker panel. Finally, remove those welds along the taillight and rear valance area and the quarter panel should lift off. Of course, you won't force it until you find out why it isn't moving.

Panel Discussion—Normally, you should now be able to put on the new quarter pan-

Damaged part of inner fender is cut away. On convertibles, it's difficult to drill out all spot welds holding inner fender to body. Therefore, Larry will section-in the inner fender.

New inner fender is set in place and scribed to match cut-away portion. After cutting on scribe line, inner fender is clamped into place, ready to be welded. First, tack weld fender about every 2 in. Then, weld seam using as little heat as possible. This will help prevent warpage.

Plug-weld and braze new panel where old spot welds were made. Finish job by filling with lead or plastic.

el. On our project car, this was not the case. The inner fender had been destroyed in the process of flaring the fenders. Now, Larry has to section in a piece of inner fender so that the quarter panel will have something to attach to.

If this were a fastback or notchback, Larry would simply drill out the spot welds along the inner fender just as he did for the quarter panel. On the convertible however, this is a very difficult and time-consuming operation. Instead, Larry cuts away part of the inner fender, leaving that area that would be so difficult to remove. He then places the new inner fender over the old, scribes lines and cuts away the part that is unusable.

The new piece is then fitted to the old line, a little trimming is done and the part is ready to be butt-welded. Note that the part is tack-welded first. Tack each end, the center, and then about every 6 in. and finally about every 2 in.

Using the smallest oxyacetylene tip possible with the least amount of heat conducive to good penetration, Larry now butt-welds the length of the inner-panel seam. Occasionally, it is necessary to do a little hammer work and a little quenching to keep the metal from warping. If you get too much warpage, you're using too big a tip. Reduce the tip size and start again. When finished, the inner fender will be treated with undercoating, covering the weld seam.

Fitting New Quarter Panel—The new panel is raised into place and aligned with the location of the old panel. This is easily judged at the area of the trunk opening, around the taillight and the door jam. Use a long punch or Phillips-head screwdriver to pass through cut-out areas to adjust the panel to the body. By using the punch as a lever, both large and small movements may

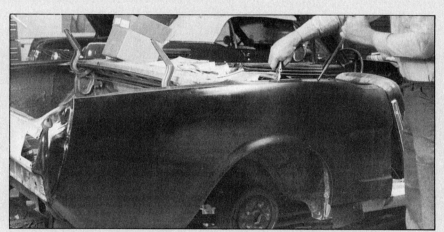

Larry uses a large drift punch in holes for top reveal molding to align quarter panel. Vise-Grip pliers clamp panel tight to body.

be made. When you have the panel aligned, clamp it down with Vise-Grip pliers, C-clamps, Cleco clamps or equivalent.

Close the trunk lid and check the fit. Open and close the door, checking its fit. Strive for an even 1/8—3/16-in. clearance all the way around; both at the door and trunk. Don't be afraid to do a little filing and grinding to get the desired fit. When you have what you think is the perfect fit, remove the panel from the car.

Install Panel—Where you drilled out the spot welds on the old panel, drill 1/4-in. holes in the new panel. These will be used to make *plug welds*. Plug, or *rosette*, welds are simply welds that fill the void created by drilling the hole. This allows you to attach two pieces of metal together that mate on their flat sides. Now, remount the panel,

adjust it for that perfect fit, clamp it tight and begin filing the plug welds.

Where the metal is thin, plug welds may cause warpage. In this case, brazing is quite acceptable. Brazing is also acceptable where the panel meets the rocker. Later, the seam line may be reformed with the edge of a file. On the convertible, edge-weld the seam around the top well. This will later be covered by a chrome reveal molding, so elegant welds are not necessary.

Finish the job by grinding off any high spots on the welding or brazing. Then lead the seam at the roof line, the door frame, trunk opening, and along the radius of the fender well. Clean the leaded area with an acid neutralizer and then spray on some primer. The job is complete now and ready for paint.

diagonal strokes. Cut a few strokes top to bottom, right to left. Then change to a left-to-right stroke, producing a diamond design. Move along the panel so that you do not wear through in one place.

After a few dozen strokes, you'll see that most of the black paint in the guide coat has been removed, leaving only a few areas where it can still be seen. These are the low areas that remain. There are two ways to raise them. If you've applied enough surfacer, keep sanding the high spots until they are level with the low spots. The other way is to use one of the other fillers.

If the valley is not too deep, apply a coat of glazing putty, either lacquer-based or polyester. When this is dry, sand it down to match the surrounding area. Wrap sandpaper around a small, flat board (I use a piece of paint-stirring stick) and sand only inside the putty area. If you work outside the putty area, you'll dig into the already leveled surfacer. The result will be a visible ring under the paint. Therefore, be very careful when sanding the glazing putty to avoid working into the surrounding area. When the putty is level with the other surfaces (you can feel it with your hand), finish off with the sanding block. This will make everything even.

Go back now and spray-on another coat of surfacer and repeat the process. This time, no black spots should remain. If there are any, fill them. Pin holes should be handled with spot putty rather than glazing putty.

TYPES OF PAINTS

There are many types of paints to choose from. Each has been formulated to meet some specific purpose and to give a particular result. To determine the type of paint to use, ask yourself a few questions: Where am I going to paint the car? Will it be at home? Will I use a paint booth? What kind of equipment do I have at my disposal? The answers to these questions will determine the paint you use. The following list of paints will help explain how:

Acrylic Lacquer—Acrylic lacquer has been the choice of the home craftsman for many years. It is a fast-drying paint that, when applied correctly, flows well and covers easily. A fast-drying paint prohibits the adhering of dust to the fresh paint when applied in a "contaminated" area such as a garage or driveway. Unfortunately, lacquer has very little natural gloss. The gloss must be achieved by sanding with ultrafine (#600—1000-grit) sandpaper, followed by rubbing compound. This produces what some consider to be the most beautiful of

Careful masking—both inside and out—is an important part of painting, especially if parts to be painted remain on car. Photo by Michael Lutfy, courtesy of *Mustang Illustrated* magazine.

all paint jobs. It develops a luster of its own, rather than the "wet" look of the enamels. Consider using lacquer if you are painting your car in a driveway or home garage.

Synthetic Enamel—Synthetic enamel is the choice of production shops with drying ovens. It is the least expensive of the five types of paint and dries to a beautiful shine when baked at a temperature of 140—200F. Use this paint only if you have access to a good spray booth and drying oven. If there is no oven, plan to leave your car in a dust-free booth for at least 24 hours.

Acrylic Enamel—Acrylic enamel dries faster than synthetic, with a higher gloss and a longer life. It fades and chalks much less than synthetic enamel or lacquer. Likewise, it also gives a tougher finish with less tendency to chip, crack and peel. And, most importantly to you restorers, it is what Ford used when it built your Mustang.

When catalyzed with hardener, it can be buffed the same as lacquer. This is a very expensive way to do the job, but gives the most beautiful results for the craftsman working at home.

Base Coat/Clear Coat—If you like that ultra-high gloss and super-wet look, the base-coat/clear-coat system is the way to go. A color coat of acrylic enamel covers the car and gives it color. A catalyzed clear coat is then applied to give a gloss. This system is best done in a booth where dust cannot get to the paint. If a booth is not

possible, the clear coat can be buffed to remove any surface dust or dirt. The buffing imparts a gloss of its own. Some of the wet look is sacrificed in this operation, but the advantages far outweigh the gloss loss.

These are the paints best suited to the home craftsman. There are others, of course, such as the urethanes. They are designed more for commercial and industrial use where chemical or environmental damage might be a problem. Lacquer, synthetic enamel, acrylic enamel and catalyzed acrylic enamel best serve the major portion of the industry.

PREPPING

Prepping a panel simply means using an acid etch to clean it and give a "tooth" to which the primer can cling. Use one of the available metal preps, such as DuPont or Ditzler. Mix it with water according to directions. Wearing rubber gloves and, using very coarse steel wool, apply metal prep to the bare metal and scrub with steel wool. When all remaining paint, stains and rust are gone, wipe off the metal prep with a clean, dry rag or towel.

The directions don't suggest it, but I like to then wipe everything down with a generous wash of lacquer thinner. This removes any acid you might have missed with the dry towel. After this final operation and when you're sure all surfaces are thoroughly dry, apply a coat of primer. You can do bodywork over primer just fine. If you

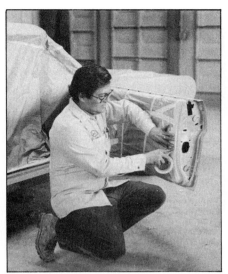

Mask off inside of door prior to painting door jams. Overspray here and in engine compartment is a sign of poor work.

Apply sealer coat before color coat to minimize sanding scratches showing through. Photo by Michael Lutfy, courtesy of *Mustang Illustrated* magazine.

choose to do leadwork, you'll have to grind the primer away to bright metal for the lead to adhere.

PAINTING EXTERIOR

You should now be ready to paint the car. The bodywork is complete, everything has been block-sanded until each panel is perfect. A paint system has been selected and everything is "go."

If you wish to paint the car the way it was done at the factory, hood and fenders must be removed from the car. This means the headlight buckets and headlight doors must come off. For the best job possible, everything will be disassembled as described earlier. This gets paint to all surfaces of the metal, preventing rust problems down the line.

Or, you can paint the car with the hood, fenders, doors, trunklid and headlight doors installed after first *cutting in* those sections that might not recieve a full measure of paint. That means pre-painting door jams, the underside and edges of the hood and trunklid and headlight doors, then installing these parts. In most cases, small add-ons like sport mirrors, scoops and spoilers should be painted *off* the car. This is the way many restoration shops do the job and helps eliminate any possibility of panel-to-panel tonal mismatch—all too common with metallics.

Once again, coat all surfaces with another medium coat of surfacer. Then, hand-sand with #400 wet or #320 dry paper.

This will cover sanding scratches caused by block-sanding. The following operations should be carried out in as clean an area as possible.

If this is to be a garage paint job, sweep the floor as carefully as possible. Cover everything that could be damaged by overspray such as washing machines, driers and other appliances. *Turn off any pilot lights!* Although paint fumes are not as explosive as gasoline fumes, they are in themselves, highly flammable. Don't trust to luck. She may not be a lady in your case. A few minutes before you start the actual painting, wet-down the floor and walls as much as is practical. A garden sprayer with the nozzle set on fine works well.

If you're working in a regular spray booth, follow the same procedures. Sweep well, allow the dust to settle and then wet the floors to keep the dust down.

Remember, it is very important to paint the entire car at one time, rather than a piece now and another part next week. This is particularly true with metallics. Changes in air pressure, the settling of metallics in the can, temperature and humidity will work to your disadvantage. The result will be a difference in color and tone between the parts that were painted at different times. Don't risk a "funny looking" fender. Paint all parts at one time.

Sealer—Although it will add some expense to your paint job, I recommend applying sealer before the color coat. A

non-sanding sealer cuts down on the number of sanding scratches showing through. Second, it seals everything underneath, which prevents anything from coming through. And last, amateur painters will often apply the first coat of paint too wet. This causes sanding scratches to swell. The use of a sealer prevents this.

Read the directions on the sealer carefully. Of course, you are using a system so the sealer and paint are compatible. Sealers have a different flash time, however, among different brands. Determine the amount of time to wait between the sealer coat and the first coat of color. Follow these instructions carefully, lest you ruin the job at this late point.

Three-Coat Enamel System—Whatever my instructions are in the following paragraphs, disregard them if they do not coincide with the instructions on the paint can. Read those instructions carefully. If you do not understand them or have any questions about them, discuss it with the paint dealer or factory rep. Don't go off half-cocked at this point; you've come too far.

Most painters apply three coats of paint. The first is a light coat called the *tack coat*. This gives the second coat, or color coat, something to stick to and allows it to be applied a little heavier without running. The third and final coat is applied quite heavily, just short of creating runs. This is called the *gloss coat* and gives enamel its depth and shine. Lacquers work a little differently.

Lacquer—Lacquer is built up by applying one medium coat on top of another until the paint has been built to a depth that will accept sanding and buffing without creating thin spots. This is generally five or six coats. Too many coats and the lacquer becomes brittle and prone to cracking. To prevent cracking, some painters add an ounce or two of flex agent to the paint to reduce the brittle effect.

Flex agent, when used in large quantities, makes lacquer flexible enough to use on vinyls and other soft plastics. Another trick used by the professionals when applying lacquer is to use a very slow-drying thinner on the last coat. This allows the lacquer to flow out better, eliminating the rough texture—orange peel—that is characteristic of quick-drying lacquer.

Paint Application—After going over all parts of the car with a tack rag, begin the paint job by painting the underside of the hood and trunk, then inside the fenders. While these are drying, paint the floorpan and inside rear quarter panels, including the area behind the rear seat. Move out and

Don't forget to recaulk seams, like this one at door-hinge-post-to-cowl joint. Cracked, dried old sealer can be scraped out with screwdriver. Photo by Tom Monroe.

Undercoat rust-prone, lower sections of fenders and quarter panels. Do it now while access is easy. Note tape over fender-script attaching holes to keep undercoating from spraying through onto exterior. Photo by Ron Sessions.

Interior panels can be spruced up with semi-gloss paint or, in the case of plastic panels, dye.

do the inside of the trunk. When the hood and trunk are dry enough to be moved, turn them over and prepare to paint the top sides along with the rest of the car.

Be sure the engine compartment has been masked off properly or you'll get overspray in there when you paint the cowl. Start the main body of the paint job by doing the door jams and top of the rockers first. If the car is a notchback, paint the top first then the cowl, the trunk lid and rear of the car, then the rear quarter panels. Move to the doors, hood, valances and other small pieces.

This is not the only pattern to use. It is only one of many. If you choose to do it another way, just be sure to plan ahead and not drag the air hose over fresh paint.

I have only been able to hit the high points of painting your Mustang. Again, I suggest that you read HPBooks' *Paint & Body Handbook* for a more in-depth treatment of the subject.

PAINTING INTERIOR

In most cases, the interior of the car must also be painted. This includes the instrument panel, steering column, inside of the doors around the door panel and, in early models, the rear quarters.

Carefully mask any areas of the exterior that might receive overspray. This is especially true for the interior doors and the instrument panel where it meets the cowl.

At the factory, interior parts were painted a semi-gloss to medium-flat finish. The best way to describe it is by discussing the two types of paint that will achieve the desired effect.

Lacquer—One way to approach the problem is to use lacquer with a medium thinner. On the exterior, I mentioned the use of a slow thinner that allows the paint to flow out more smoothly. But the opposite is desired for interior parts because you want that dull, velvety "orange peel" effect. If you use a medium reducer with the lacquer, it should give the desired finish.

Enamel—The second approach is to use an acrylic enamel with a little flattening agent. Here, I would experiment with the amount of flattener to use. Reduce the enamel according to directions. To one quart of reduced material add two ounces of flattening agent. Spray this out and allow it to dry until the gloss disappears. See if this provides the desired effect.

You do not want it to be so flat that it looks like wrought iron. The effect the factory wanted was to look like vinyl upholstery. That's the reason for the graining effect stamped into the inside-door and inside-rear-quarter panels on '65—68 models. When the finish has a soft, velvety appearance with just a touch of shine, you've found the right combination of paint and flattener. Now go ahead and spray the surfaces. This time, however, omit the third step for the gloss coat.

In the next chapter, you'll begin to reassemble your car. It looks pretty good now. Just wait until it's complete!

EXTERIOR COLOR CODES

1964-1/2

Code	Color	Paint	Type
A or 9A	Raven Black	RM A946	Lacquer
		RM P403	Enamel
B or 2B	Pagoda Green	RM 1628	
C or 6C	Special White	RM 2347	
F or 3F	Guardsman Blue	RM 1630	
H or 5H	Caspian Blue	RM 1770	
J, 3J, J9	Rangoon Red	RM 1440	
K or 3K	Silversmoke Gray	RM 1632	
M or 5M	Wimbledon White	RM 1683	
P or 4P	Prairie Bronze	RM 1693	
V	Sunlight Yellow	RM 1692	
X	Vintage Burgundy	RM 1636	
Y	Skylight Blue	RM 1637	
Z	Chantilly Beige	Ditzler 22393	
3 or 3B	Poppy Red	RM 1774	
5 or 5A	Twilight Turquoise	RM 1736	
7	Phoenician Yellow	RM 1634	

1965

Code	Color	Paint	Type
A or 9A	Raven Black	RM A946	Lacquer
		RM P403	Enamel
B or 2B	Pagoda Green	RM 1628	
C or 6C	Honey Gold	RM 1739	
D or 3D	Dynasty Green	RM 1629	
F or 3F	Arcadian Blue	RM 1223 (Late '65)	
H or 5H	Caspian Blue	RM 1770	
I	Champagne Beige	RM 1735	
J, 3J J9	Rangoon Red	RM 1440	
K or 3K	Silversmoke Gray	RM 1632	
M or 5M	Wimbledon White	RM 1633	
O	Tropical Turquoise	RM 1737	
P or 4P	Prairie Bronze	RM 1693	
R	Ivy Green	RM 1738	
V	Sunlight Yellow	RM 1692	
X	Vintage Burgundy	RM 1636	
Y	Silver Blue	RM 1734	
3 or 3B	Poppy Red	RM 1774	
8	Springtime Yellow	RM 1776	

1966

Code	Color	Paint	Type
A or 9A	Raven Black	RM A946	Lacquer
		RM P403	Enamel
F or 3F	Arcadian Blue	RM 1223	
H or 5H	Sahara Beige	RM 1779	
K or 3K	Nightmist Blue	RM 1780	
M or 5M	Wimbledon White	RM 1633	
P or 4P	Antique Bronze	RM 1781	
Q or 4Q	Brittany Blue	Dupont 4951	
R	Dark Green Metallic	RM 1738	
T or 5T	Candyapple Red	RM 1782	
U or U9	Tahoe Turquoise	RM 1231	
V	Emberglo	RM 1777	
X	Vintage Burgundy	RM 1636	
Y	Silver Blue	RM 1734	
4 or 4F	Silver Frost	RM 1784	
5 or 5A	Signalflare Red	RM 1791	
8	Springtime Yellow	RM 1776	

1967

Code	Color	Paint	Type
A or 9A	Raven Black	RM A946	Lacquer
		RM P403	Enamel
B or 2B	Frosty Turquoise	RM 1741	
D or 3D	Acapulco Blue	RM 1935	
F or 3F	Arcadian Blue	RM 1223	
H or 5H	Diamond Green (Shown as Lincoln/T-Bird Only)		
I	Lime Green	RM 1882	
K or 3K	Nightmist Blue	RM 1780	

Code	Color	Paint	Type
M or 5M	Wimbledon White	RM 1633	
Q or 4Q	Brittany Blue	RM 1643	
T or 5T	Candyapple Red	RM 1782	
V	Burnt Amber	RM 1881	
W	Clearwater Aqua	RM 1880	
X	Vintage Burgundy	RM 1636	
Y	Dark Moss Green	RM 1879	
Z or Z9	Sauterne Gold	RM 1783	
4 or 4F	Silver Frost	RM 1784	
6 or 6E	Pebble Beige	RM 1635	
8	Springtime Yellow	RM 1776	

1968

Code	Color	Paint	Type
A or 9A	Raven Black	RM A946	Lacquer
		RM P403	Enamel
B or 2B	Royal Maroon	RM 1941	
D or 3D	Acapulco Blue	RM 1935	
F or 3F	Gulfstream Aqua	RM 1942	
I	Lime Gold	RM 1882	
M or 5M	Wimbledon White	RM 1633	
N or 4N	Seafoam Green	RM 1885	
Q or 4Q	Brittany Blue	RM 1643	
R	Highland Green	RM 1946	
T or 5T	Candyapple Red	RM 1782	
U or U9	Tahoe Turquoise	RM 1231	
W	Meadowlane Yellow	RM 1949	
X	Presidential Blue	RM 1950	
Y	Sunlit Gold	RM 1951	
6 or 6E	Pebble Beige	RM 1635	

1969

Code	Color	Paint	Type
A or 9A	Raven Black	RM A946	Lacquer
		RM P403	Enamel
B or 2B	Royal Maroon	RM 1941	
C or 6C	Black Jade	RM 2038	
E or 4E	Aztec Aqua	RM 2036	
F or 3F	Gulfstream Aqua	RM 1942	
I	Lime Gold	RM 1882	
K or 3K	Bright Gold Metallic	RM 2198	
M or 5M	Wimbledon White	RM 1633	
P or 4P	Winter Blue	RM 2035	
Q or 4Q	Brittany Blue	RM 1643	
S	Champagne Gold	RM 2044	
T or 5T	Candyapple Red	RM 1782	
W	Meadowlark Yellow	RM 1949	
Y	Indian Fire	RM 2045	
2 or 2J	Lime	RM 2037	
3 or 3B	Calypso Coral	RM 2197	
4 or 4F	Silver Jade	RM 2040	
6 or 6E	Pastel Gray	RM 2033	

1970

Code	Color	Paint	Type
A or 9A	Raven Black	RM A946	Lacquer
		RM P403	Enamel
C or 6C	Dark Ivy Green Metallic	RM 2308	
D or 3D	Yellow	RM 2309	
G	Medium Lime Metallic	RM 2311	
J, 3J, J9	Grabber Blue	RM 2313	
K or 3K	Bright Gold Metallic	RM 2198	
M or 5M	Wimbledon White	RM 1633	
N or 4N	Pastel Blue	RM 1238	
Q or 4Q	Medium Blue Metallic	RM 2316	
S	Medium Gold Metallic	RM 2044	
T or 5T	Red	RM 1782	
U or U9	Grabber Orange	RM 2318	
Z or Z9	Grabber Green	RM 2328	
1	Calypso Coral	RM 1734	
2 or 2J	Light Ivy Yellow	RM 2037	

9

Assembly

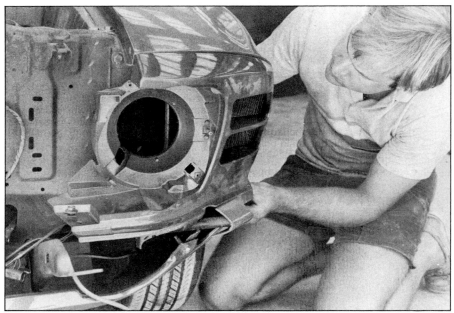

Dave Stroot is assembling his 1970 convertible after painting it.

With engine rebuilt and installed in freshened up engine compartment, car is beginning to take shape.

Now it's time to put your Mustang back together. You'll be thankful you labeled everything, took photos, drew diagrams and put everything away for safe keeping. Start with the engine and engine compartment. With the fenders off, it's easy to work on these parts, and you don't run the risk of scratching new paint. Leave the fenders off until you're sure everything mechanical that's accessible from up top is working correctly.

When the engine and engine compartment are complete, we'll wire the car. Then, you can start the engine, move the car around a little and be sure everything works. When you're sure the car's a runner, you can bolt on the sheet metal.

This chapter concludes with a section on glass. You can install the windshield on the convertible along with the side and quarter windows. On the fastback and notchback, leave the windshield and backlite out until you've installed the headliner. It actually fits *under* the glass, both front and rear. However, for the sake of organization, I explain how the windshield and backlite are installed. I know you're anxious; so let's get on with it.

ENGINE, TRANSMISSION & DRIVE LINE

At this point your engine should be sitting in the engine compartment, mounted to the transmission and engine mounts. Once again, elevate the car so you can roll under it with your creeper. *Be sure to set the jack stands.*

At the end of this chapter is a list of fastener torque specifications you'll need as you assemble components to the engine. Refer to these as you tighten things. Nuts and bolts that are not secured to specification will work loose. Overtightening will chance stripping the threads or actually breaking the bolt or stud. Pay careful attention to torque limits.

Drive Shaft—Inspect the slip yoke for rust. Clean it of old grease and any rust and apply a new coat of grease to the yoke splines. Install new universal joints at both ends of the drive shaft. If the lip seal at the end of the transmission extension housing is worn or shows signs of leakage, replace it.

Slide the slip yoke onto the transmission output shaft, install the rear of the drive shaft onto the pinion flange and secure it with two U-bolts. Torque to specification. Make sure you follow your alignment marks when connecting the drive shaft to the differential. If you replaced the shaft, be alert for drive-shaft vibration during the first test drive. If any are felt, disconnect the drive shaft from the axle flange and rotate it 180°. Sometimes, a sympathetic vibration occurs between the shaft and axle-pinion flange. Rotating the drive shaft 180° may neutralize this problem. If it doesn't, check the shaft for binds, and have

If you remembered to mark everything during disassembly, putting it back together correctly will be a snap. Here, driveshaft flanges were marked to maintain factory balance.

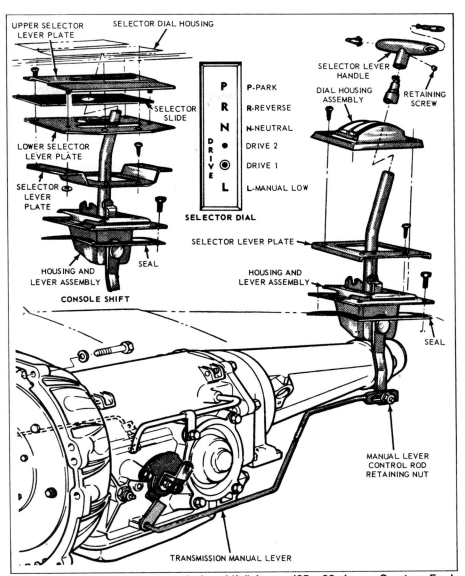

Typical Mustang automatic-transmission shift linkage—'65—66 shown. Courtesy Ford Motor Company.

it balanced by a transmission shop.

Shock Absorbers—There should now be enough weight on the front and rear springs to allow you to install new shock absorbers. Again, replace these as new. Use a good, heavy-duty shock to ensure a comfortable ride and long tire wear.

There is a wide variety of shock absorbers available to the automotive enthusiast. Original-equipment manufacturer (OEM) shocks are standard for an all-original restoration. If, however, you are restoring your car for performance, consider a heavy-duty, aftermarket shock absorber. There are several types to choose from.

Conventional—Heavy-duty conventional shocks are similar to the OEM. They use hydraulic fluid as the damping medium, metered through a series of orifices and valves. As it implies, a heavy-duty shock has a larger hydraulic chamber, better seals and, consequently, longer life.

Air Shocks—At one time, air shocks were popular performance add-ons. Because air shocks increase damping rate when compressed air is added, air shocks provided added traction for drag racing. In effect, the rear of the car acts like it has higher-rate springs in it, subduing axle tramp during hard acceleration. Air shocks also raise the car when air is added, giving the proper stink-bug rake popular at the strip.

But air shocks are also used for keeping the car level when extra weight is added. If you routinely haul heavy weights in the trunk—newspapers for the Cub Scout recycling drive, haybales or the toolbox, or if you tow a trailer, then air shocks are the

best way to level the car. A quick shot of compressed air will raise the back of the car, keeping the headlights correctly aimed and maintaining correct ride height.

The trouble with air shocks is that most offer no increased performance over stock shocks when they have no air in them. Pumped up, they stiffen the ride considerably and overdamp the suspension. They are mainly for weight compensation or axle control under special circumstances.

Gas Shocks—The terms air and gas might cause confusion at first, but after trying both, you'll remember gas shocks. Gas

shocks use nitrogen gas sealed in the shock to keep constant pressure on the shock fluid at all times. This lessens the chance of fluid aeration, which causes the shock to cavitate under hard use. Gas shocks are popular now and getting more so because they give consistent damping control and increased ride comfort. They do not stiffen the ride like air shocks.

Unlike air shocks, gas shocks are sealed units. No air is added, ever, nor do gas shocks increase ride height. There are no lines to plumb or valves to install as with air shocks.

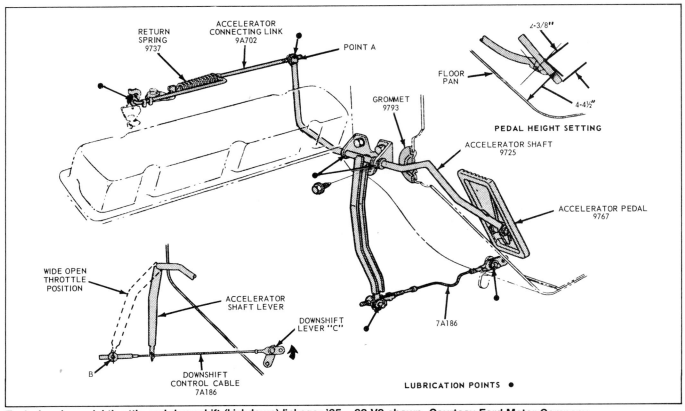

Typical early-model throttle and downshift (kickdown) linkage: '65—68 V8 shown. Courtesy Ford Motor Company.

Many aftermarket shocks are supplied in various colors, or have stickers on them. If you don't want anyone to notice what you have under the car, a few passes with black spray paint will fool all but the most determined show judge.

Urethane Bushings—We talked about urethane bushings in the suspension chapter, but as a reminder, urethane bushings will give improved suspension and steering response. Urethane bushings provide improved response because they are less compliant than rubber parts, yet do provide an adequate measure of isolation. They do not affect ride harshness, and get the most out of modern shocks and tires.

Transmission Linkage—If you removed the transmission selector lever, now is the time to replace it. Install the selector lever into the housing, securing it with its attaching nut. Torque to 20—25 ft-lb. Install the dial-indicator light. If you removed the wiring harness, the light cannot be connected until a harness has been installed. Make a note to connect the gear-indicator light into the harness.

Next, install the selector-lever handle so that you can adjust the selector-lever detent pawl. Push the selector lever all the way forward and check the clearance between the detent pawl and plate. The clearance should be 0.005—0.010 in. Adjust clearance by rotating the detent-adjustment screw right or left.

Remove the handle and finish assembling the unit by installing the selector-lever plate, dial-housing assembly, and if in a console-equipped auto, the cover plate. Finish by installing the selector-lever handle. If you don't have the console in yet, install the shift lever, then the console. Install the console, then the right seat and the shift knob or handle.

Place the selector lever in the D position. Scoot under the car and install the manual-lever rod to the transmission. After the car is running, adjust this linkage.

Before turning your attention away from the transmission, the downshift control must be hooked up. This requires that the throttle linkage be assembled.

V8 Throttle Linkage—Install a new accelerator-shaft grommet in the dash panel (firewall) and push the accelerator shaft through this grommet from outside the firewall. Mount the accelerator-shaft lever to the firewall with its sheet-metal screws. Connect the accelerator shaft to the accelerator connecting link coming from the carburetor. Attach the accelerator pedal to the other end of the accelerator shaft inside the car.

Attach the throttle-return spring, then have a helper step on the accelerator pedal while you peer into the carburetor. Manually prop the choke plate open and check the throttle plates. They should be wide open. Also try the linkage at the carburetor. There should be no slack in the linkage.

On '69—70 cars with throttle cables, check for wide-open throttle. If you can't get the throttle plates to open fully, check for a stretched cable. Have your helper continue to floor the accelerator while you work the linkage at the carburetor. If you can get the throttle to open more by pulling on the linkage, then the cable is stretching. Replace it. Binding or fraying is also cause for cable replacement. There is no adjustment on cable-type throttle linkage.

Because you are reinstalling parts that were originally on the car, there should be no problem with the adjustment. If there is, you can probably handle 90% of the adjustments necessary by shortening or lengthening the throttle-linkage rod right before the carburetor on pre-'69 cars. If not, check the shop manual for your car. Extensive instructions for setting up the linkage from scratch are given there.

Six-Cylinder Throttle Linkage— Installing the throttle linkage on a six-cylinder Mustang and adjusting the downshift cable or rod is similar to V8 models with the following exceptions:

• An accelerator-shaft damper is added between the accelerator shaft and firewall next to the accelerator-shaft lever.

• The accelerator connecting link attaches to the bellcrank on the stabilizer bracket.

• A stabilizer rod returns to a second stabilizer bracket on the dash panel. This is the point where the adjustment is made for accelerator-pedal height (4 1/2 in.)

Automatic-Transmission Kickdown— Now that the throttle is set, disconnect the kickdown rod or cable and throttle-linkage return springs at the carburetor. Open the throttle completely using the throttle linkage. A helper or weight on the throttle is best. Also hold the transmission kickdown rod or cable in its full downshift position—back or down depending on how you visualize kickdown movement. Turn the screw on the carb kickdown lever to within 0.040—0.080 in. of contacting the carburetor-throttle lever. Release the throttle linkage, kickdown rod or cable and connect the return springs.

To install the automatic-transmission kickdown linkage on '68-and-earlier cars, make sure the cable is detached from the forward end of the kickdown bellcrank. The kickdown bellcrank is mounted vertically, running from the carburetor to the transmission. Making sure the carburetor choke is completely open, press the accelerator pedal all the way to the floor. A helper or heavy weight can do this while you work in the engine compartment. At the other end of the cable, at the transmission, there is a lever. Rotate it all the way forward. While holding it forward, pull the cable forward also, removing all the slack. Watch where the cable end is in relation to the kickdown bellcrank. Adjust the cable so it will just fit in the bellcrank, then turn the adjustment one turn tighter. This will ensure the transmission kicks down when

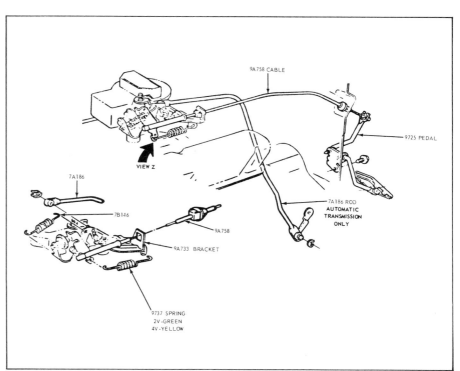

Typical late-model throttle and downshift (kickdown) linkage: '69—70 V8 shown. Courtesy Ford Motor Company.

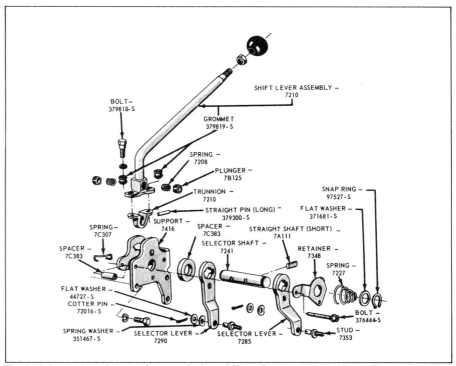

Typical three-speed manual-transmission shifter. Courtesy Ford Motor Company.

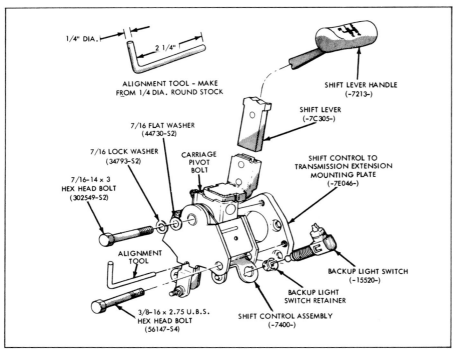

Hurst supplied four-speed shifter: '69—70: Note alignment tool. It must slide in and out for correct linkage adjustment. Courtesy Ford Motor Company.

the throttle is opened. Now, connect the cable to the bellcrank.

On '69—70 six-cylinder cars, the procedure is the same, but there is no cable. Instead, the kickdown rod connects directly to the lever on the transmission. Remove the throttle return spring and undo the kickdown rod from the transmission lever. Open the throttle completely using the accelerator pedal, and hold the transmission lever all the way forward. Adjust the rod so it aligns perfectly with the transmission lever, then turn it one turn tighter. Connect the rod to the transmission lever, the install the throttle return spring.

Manual Transmission—Connect the shift linkage from the shift lever to the transmission by positioning the shift-selector assembly to the extension housing and install the attaching bolts. On the four-speed manual transmission, there are three selector levers rather than two as on the three-speed transmission.

Transmission Final Assembly—Route the speedometer cable through the dash panel and connect to the transmission. Connect the backup-light switch.

On the automatic transmission, install the filler tube, connect the neutral-start switch and install the fluid-cooler lines. These may be connected at the other end when the radiator is installed.

Clutch—Begin by installing the equalizer bar, using new bushings. To install the equalizer bar, position the equalizer bar, washers, inner bushing and retainer on the inner stud. Holding the equalizer bar in place, install the outer bracket. Connect the release rod and spring to the bottom of the equalizer bar. Finally, connect the clutch-pedal rod and spring from the equalizer bar to the clutch-pedal lever. Torque to specifications.

The drive line is now complete. Turn your attention to the exhaust system.

Exhaust System—Mount the exhaust manifold(s) to the head(s) with new metal-clad gasket(s). Tighten the nuts and bolts to specification.

Assemble the muffler and outlet pipe and loosely mount them under the car. If it helps, wire the outlet side up until everything is assembled.

On convertibles, be sure the front-floor lower-crossmember plate has been removed. Run the muffler-inlet pipe from the exhaust manifold to the muffler and again, suspend it in place with wire.

Install a new gasket on the inlet-pipe flange. Connect the inlet pipe to the exhaust manifold and muffler and attach the stud nuts at the flange. On big-blocks, don't forget the heat-riser valve. Position the clamp at the muffler. Do not tighten the

flange nuts or muffler clamp.

Mount the inlet-pipe bracket to the side of the engine block and clamp the inlet pipe to this bracket. Leave the clamp loose.

Remove any wires supporting the system. Adjust the components so there is no contact between the exhaust and the body or chassis. Metal-to-metal contact may cause an engine-speed-sensitive vibration called *grounding*. Working from the front to the rear, tighten all clamps and nuts to specification. On convertibles, replace the front-floor lower-crossmember plate.

Power-Steering-Pump Pulley—Having rebuilt your power-steering pump back in chapter 5, it's now time to mount it to the engine and connect the hoses. But first, you must mount the pulley to the pump. *Do not force the pulley onto the shaft with a hammer!* This will damage the internal thrust surfaces.

If you have access to Ford tool T65P-3A733-A or equivalent, use it to install the power-steering-pump pulley. Or you can make your own pulley installation tool. You'll need a 3/8-16 (coarse-thread) hex-head bolt. The bolt should be threaded all the way down the shank. This is a hard number to find at the local hardware store, so you may have to settle for a carriage bolt. A carriage bolt has a round head. It will serve your purpose, however, if you file two flat spots on the head so that it can be securely clamped into a vise. Buy a nut and two large fender washers.

To use this as a pulley installation tool, thread the nut onto the bolt all the way to the head. Then, add the two washers. Spread bearing grease around the shaft hole in the pulley and on the power-steering shaft. Slip the pulley over the bolt and let the face rest against the washers.

Thread the bolt into the power-steering-pump shaft until it seats and begins to turn the shaft. Now, clamp the bolt head into the vise.

You can see what comes next. Run the nut up, pushing the pulley onto the shaft. Stop when the pulley is flush with the end of the shaft. Remove the "tool" and insert a 3/8-16-in. bolt and washer into the shaft to retain the pulley. This needs to be snugged down tight. Do not force it, however.

Pump Installation (Six-Cylinder & 289-CID V8)—With the pump on the bench, install the mounting bracket and torque the bolts to specification. Set the pump and bracket up to the engine and install the attaching bolts. On the six-cylinder, one bolt will attach the brace to the bottom of the bracket, and the other end of the brace is attached to the block. On the V8, three bolts retain the bracket and there is no

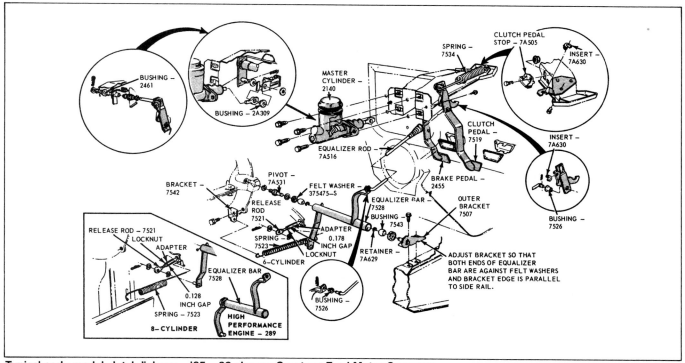

Typical early-model clutch linkage—'65—66 shown. Courtesy Ford Motor Company.

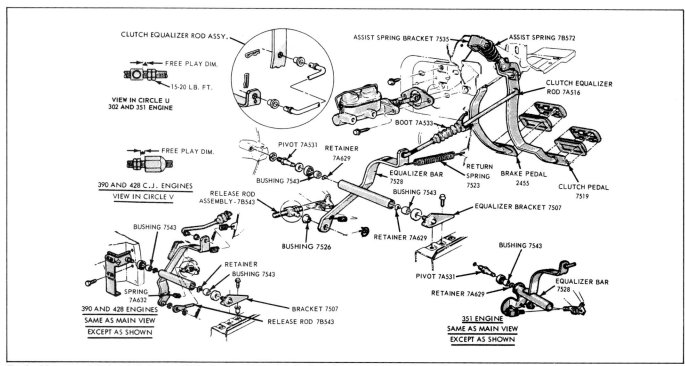

Typical late-model clutch linkage—'70 shown, '67—68 similar. Courtesy Ford Motor Company.

Jury-rigged device is used to draw power-steering pulley onto pump shaft. If you use a hammer to drive it on, you'll ruin pump.

It's always best to replace flexible brake hoses at restoration time. Photo by Ron Sessions.

brace. Torque all bolts to specification. Install a new pulley belt and tighten until there is less than 1/2-in. deflection when you press the belt midway of its longest run.

Before attaching the pressure hose to the rear of the pump, torque the outlet fitting to specification. Install a *new* pressure hose to the outlet fitting. Don't risk a blow-out!

Pump Installation (Big-Block V8)— There are two mounting brackets and an oil cooler on big-block V8 and late-model steering pumps. The mounting brackets consist of a casting to which the pump is mounted; then a stamped-steel bracket that retains the cast bracket to the engine block.

With the pump on the bench, mount the cast bracket to the pump with one bolt on the narrow leg. Mount the stamped bracket to the front of the cast bracket with a bolt through the adjustment slot. On the rear of the pump, fasten the rear brace to the high-pressure fitting. Align the hole at the other end of the brace with the mating hole in the cast bracket.

With the long stud threaded into the left head, slide on the spacer, the pump on its bracket (be sure the rear brace from the high-pressure fitting is also slipped over the stud), and retain them there with a nut. Fasten the stamped bracket to the engine block with two retaining bolts. Torque these two bolts to specification. Mount the pulley belt and adjust the tension to less than a 1/2-in. deflection in the middle of the longest run. Tighten the remaining bolts.

If your unit incorporates an oil cooler, this is installed on the low-pressure return line. Install the high-pressure line from the fitting on the rear of the pump to the control valve. Clamp one end of the return line to

the pump and route the other end to the cooler inlet. From the cooler outlet, route the other low-pressure return line to the low-pressure side of the control valve.

Control Valve— Slide a clamp over a new low-pressure return hose (the one with only one threaded fitting). Fit this to the reservoir and tighten the clamp. Route it to the control valve. Now connect the two hoses to the control valve; each has different-size male fittings entering the control valve. When the engine can be started, fill the reservoir with ATF, cycle the system and check for leaks.

Air-Conditioning Compressor & Condenser— The A/C compressor is mounted in conjunction with the power-steering pump. Begin by attaching the mounting base to the compressor. It helps to have a friend assist you because the compressor and clutch are heavy.

Mount the compressor to the left cylinder head (V8) and power-steering-pump bracket. Mount the adjustable idler bracket to the compressor bracket with three mounting bolts. If your car uses a fixed idler pulley—stands on three legs—mount it so the top leg fixes to the adjustable idler-pulley bracket and the lower leg to the water-pump bolt. A spacer goes *behind* the idler bracket. When the wiring has been installed, hook-up the magnetic-clutch lead.

The condenser is mounted in front of the radiator on the radiator-support bracket. Route the high-pressure gas line from the left side of the compressor to the top of the receiver on the side of the condenser. Run the low-pressure gas line on the right side of the compressor, back to the outlet on the evaporator. Finish the circuit by installing the high-pressure liquid line from the bottom of the receiver to the expansion valve on the evaporator.

BRAKE SYSTEM

I'll begin with the simplest system and work to the more complex. The simplest system is the early-model car with standard drum brakes front and rear and a single-piston master cylinder. But before you start, a few important reminders:

Always use a flare-nut, or tubing, wrench to loosen and tighten hydraulic brake-line fittings. The five-sided wrench gets a better grip on tubing nuts and helps prevent rounding off the slats. *Never, never replace a steel hydraulic brake line with copper tubing.* Copper cannot handle high brake-line pressures and may crack in use, causing a leak and loss of braking. Always use *new* flexible brake hoses—at the front wheels and rear-axle-to-body union. Lube

all sliding and metal-to-metal contact surfaces of drum and disc brakes with white grease, sparingly applied. When tightening a hydraulic fitting, square the pipe up to the cylinder at a right angle and push the pipe in so it seats before starting to tighten the nut. Failure to do so may cock the pipe in the cylinder, distort the flared seat and cause a leak. Finally, never remove a brake line and let the system sit for more than a short time. Moisture in the atmosphere enters the system and corrodes the inside of the pipes quickly.

After overhauling the brake system, you'll want to keep it in top shape. The best way you can do this is to change the brake fluid regularly. Brake experts say standard glycol-based fluid absorbs enough water in six months to make changing it worthwhile. While half a year may be overkill for a carefully driven street car, once a year is certainly not out of line. At the very least, change brake fluid every two years. Any more than that and you are asking for corrosion, leaks and possible premature brake failures. Even at two-year intervals, expect some corrosion after awhile.

The problem is glycol-based fluid's affinity for water—it is *hydroscopic*. The stuff easily absorbs water right out of the atmosphere, so with the brake system sealed as tight as possible, water will still get invited in through the smallest holes imaginable.

Once water gets in the brake fluid, it heats and boils at normal brake temperatures. Once boiling, the water and oxygen in it can be compressed. The pedal feels spongy, and braking is reduced. Under severe use, braking can be lost entirely.

An excellent substitute is silicone brake

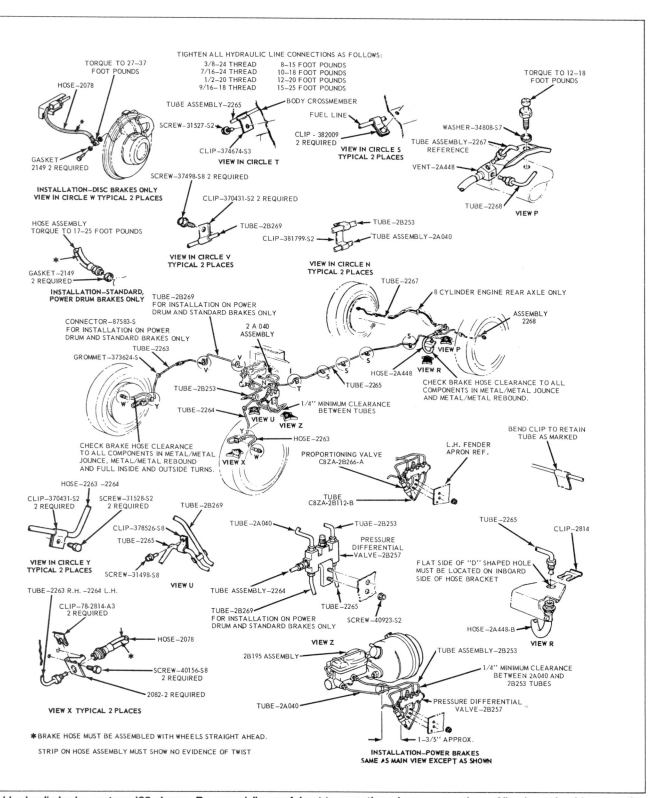

TIGHTEN ALL HYDRAULIC LINE CONNECTIONS AS FOLLOWS:

3/8–24 THREAD	8–15 FOOT POUNDS
7/16–24 THREAD	10–18 FOOT POUNDS
1/2–20 THREAD	12–20 FOOT POUNDS
9/16–18 THREAD	15–25 FOOT POUNDS

TORQUE TO 27–37 FOOT POUNDS

HOSE–2078

GASKET 2149 2 REQUIRED

INSTALLATION–DISC BRAKES ONLY VIEW IN CIRCLE W TYPICAL 2 PLACES

TUBE ASSEMBLY–2265

SCREW–31527-S2

CLIP–374674-S3

VIEW IN CIRCLE T

BODY CROSSMEMBER

FUEL LINE

CLIP - 382009 2 REQUIRED

VIEW IN CIRCLE S TYPICAL 2 PLACES

TORQUE TO 12–18 FOOT POUNDS

WASHER–34808-S7

TUBE ASSEMBLY–2267 REFERENCE

VENT–2A448

TUBE–2268

VIEW P

HOSE ASSEMBLY TORQUE TO 17–25 FOOT POUNDS

GASKET–2149 2 REQUIRED

INSTALLATION–STANDARD, POWER DRUM BRAKES ONLY

SCREW–37498-S8 2 REQUIRED

CLIP–370431-S2 2 REQUIRED

TUBE–2B269

VIEW IN CIRCLE V TYPICAL 2 PLACES

TUBE–2B253

CLIP–381799-S2

TUBE ASSEMBLY–2A040

VIEW IN CIRCLE N TYPICAL 2 PLACES

TUBE–2267

8 CYLINDER ENGINE REAR AXLE ONLY

ASSEMBLY 2268

TUBE–2B269 FOR INSTALLATION ON POWER DRUM AND STANDARD BRAKES ONLY

CONNECTOR–87583-S FOR INSTALLATION ON POWER DRUM AND STANDARD BRAKES ONLY

TUBE–2263

GROMMET–373624-S

2 A 040 ASSEMBLY

TUBE–2B253

TUBE–2264

1/4" MINIMUM CLEARANCE BETWEEN TUBES

VIEW U

VIEW Z

HOSE–2263

VIEW X

HOSE–2A448

VIEW P

VIEW R

CHECK BRAKE HOSE CLEARANCE TO ALL COMPONENTS IN METAL/METAL JOUNCE AND METAL/METAL REBOUND.

CHECK BRAKE HOSE CLEARANCE TO ALL COMPONENTS IN METAL/METAL JOUNCE, METAL/METAL REBOUND AND FULL INSIDE AND OUTSIDE TURNS.

PROPORTIONING VALVE C8ZA-2B266-A

TUBE C8ZA-2B112-B

L.H. FENDER APRON REF.

BEND CLIP TO RETAIN TUBE AS MARKED

HOSE–2263–2264

CLIP–370431-S2 2 REQUIRED

SCREW–31528-S2 2 REQUIRED

TUBE–2B269

VIEW IN CIRCLE Y TYPICAL 2 PLACES

CLIP–378526-S8

TUBE–2265

SCREW–31498-S8

VIEW U

TUBE–2A040

TUBE ASSEMBLY–2264

TUBE–2B269 FOR INSTALLATION ON POWER DRUM AND STANDARD BRAKES ONLY

TUBE–2B253

PRESSURE DIFFERENTIAL VALVE–2B257

TUBE–2265

SCREW–40923-S2

VIEW Z

TUBE–2265

CLIP–2814

FLAT SIDE OF "D" SHAPED HOLE MUST BE LOCATED ON INBOARD SIDE OF HOSE BRACKET

HOSE–2A448-B

VIEW R

TUBE–2263 R.H. –2264 L.H.

CLIP–78-2814-A3 2 REQUIRED

HOSE–2078

SCREW–40156-S8 2 REQUIRED

2082- 2 REQUIRED

VIEW X TYPICAL 2 PLACES

2B195 ASSEMBLY

TUBE ASSEMBLY–2B253

1/4" MINIMUM CLEARANCE BETWEEN 2A040 AND 2B253 TUBES

TUBE–2A040

PRESSURE DIFFERENTIAL VALVE–2B257

1-3/5" APPROX.

INSTALLATION–POWER BRAKES SAME AS MAIN VIEW EXCEPT AS SHOWN

✱BRAKE HOSE MUST BE ASSEMBLED WITH WHEELS STRAIGHT AHEAD.

STRIP ON HOSE ASSEMBLY MUST SHOW NO EVIDENCE OF TWIST

Typical hydraulic brake system: '68 shown. Be especially careful not to crossthread any connections. All unions should go together finger-tight before using flare-nut wrench. Courtesy Ford Motor Company.

fluid. Silicone fluid actually repels water, and certainly doesn't absorb it. Thus, there is no water to boil, and brake fade is practically eliminated. Water-caused corrosion is also practically eliminated. About the only way to get water in silicone brake fluid is to open the master cylinder and pour it in. You can guess the drawback—cost. Silicone fluid is much more expensive than mineral fluid, but if you figure in the reduced fluid changes and corroded brake parts, it's cheaper in the long run.

Installation—Feed the master-cylinder pushrod through the firewall. Attach the master cylinder to the firewall with two mounting bolts. Inside the car, connect the pushrod to the brake-pedal lever using a new nylon bushing.

Up through '66 models, a steel brake line passes from the front of the master cylinder to a four-way connector block directly to the left, on the inner fender. A line from the master cylinder enters the connector block in the frontmost position. On '67-and-later models with a dual-piston master cylinder, it has two inlet lines on the top of the pressure-differential valve.

The first fitting toward the front of the car on the bottom of the block is for the left-front brake. This line goes from the connector block to the flexible hose on the left brake. The outlet behind this one (bottom-rear) feeds a line to the rear brakes.

This line passes under the floorpan at the transmission tunnel and is routed alongside the fuel line. It then connects to another flexible brake hose before joining a connector block on the axle housing. Rear brake-line installation is covered in the suspension chapter, pages 66 and 67.

Finally, the rearmost outlet on the connector block on the fender apron feeds a line across the firewall to the right-front brake.

Power-Brake Booster (Early Models)—If equipped with power brakes, begin by installing the master cylinder to the booster. Next, fit the power booster to the firewall with the pushrod link passing through to the inside compartment. This is attached to the brake-pedal lever. A vacuum hose connects the check valve on the booster to the intake manifold. The rest of the installation remains the same as discussed previously.

Proportioning Valve—With front disc brakes, the system also includes a proportioning valve. It is located on the inner-fender apron below the four-way connector block or pressure-differential valve. Position the valve so that the mounting tang

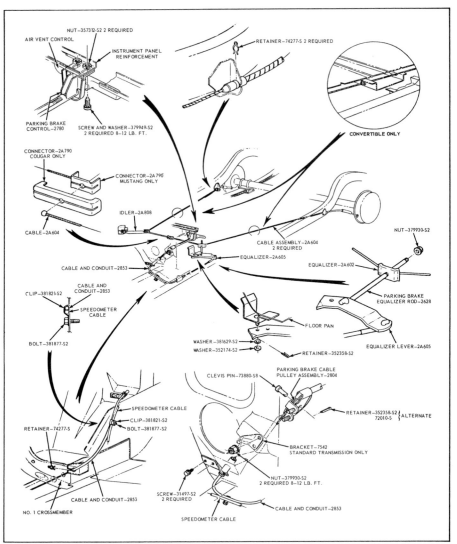

Parking-brake linkage—'65—68. Use white lithium grease wherever brake cable contacts other metal parts. Courtesy Ford Motor Company.

extends through the apron and thread-in the attaching bolt. Torque this bolt to specification.

Connect the front-to-rear brake line to the rearmost outlet on the proportioning valve. Connect the master-cylinder-to-proportioning-valve line from the top outlet on the master cylinder to the forward inlet on the proportioning valve. Torque the fittings to specification.

Pressure-Differential Valve—This valve is installed in '67-and-later models with the dual hydraulic brake system, and replaces the four-way connector block on the fender apron. The hydraulic brake-line connections for the pressure-differential valve are the same as for the connector block.

Parking Brake (1965—68)—The parking-brake-control assembly is hand-operated and consists of a control handle, bracket with locking pawl, pulley, clevis pin and clevis-pin retainer.

To assemble, lift the locking pawl, insert the control-handle rod all the way into the bracket. Connect the ball-end of the cable into the control rod. Insert the pulley and retain in place with the clevis pin. Install the clevis-pin retainer. Mount the bracket and control assembly to the firewall and instrument panel. Secure with two screws and nuts.

Parking Brake (1969—70)—The parking-brake-control assembly on later models is foot-operated. It cannot be dis-

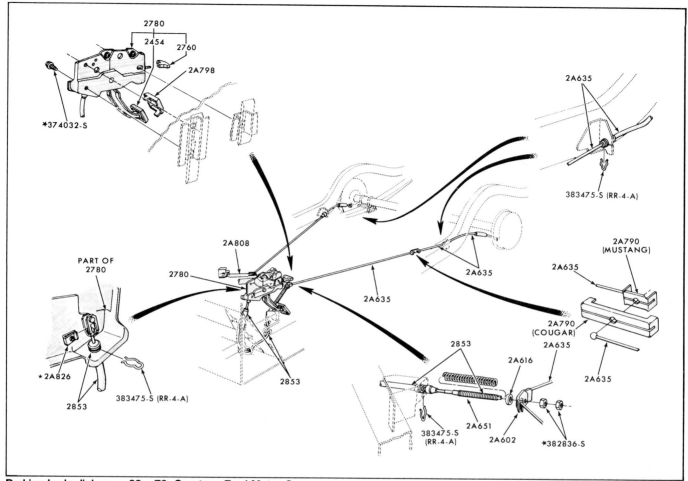

Parking-brake linkage—69—70. Courtesy Ford Motor Company.

assembled like the hand-operated units. To install the 1969—70 foot-operated units, be sure there is plenty of slack in the cable by backing off the jam nut and adjusting nut from the equalizer bar under the car as far as possible. This gives some cable slack.

Connect the ball-end of the cable to the control assembly and install the hairpin retainer. Position the control assembly on the cowl inner-side panel and install the three attaching screws. Install the parking-brake warning-light switch if the wiring harness is in the car.

Bleed the Brakes—There's more than one way to bleed a hydraulic-brake system. The two most common are *pressure bleeding* and *manual bleeding*. Pressure bleeding, because of high equipment cost, is reserved for the professional or "well-healed" hobbyist. So, if you have a pressure bleeder, you should know how to use it. I confine my discussion to manual bleeding only. If you want to learn about pressure bleeding,

pick up a copy of HPBooks' *Brake Handbook*. It will tell you everything you can imagine about brakes. With that said, let's get on with how most of us bleed brakes:

Manual bleeding requires the participation of at least two, preferably three people; one operates the brake pedal and one opens and closes the bleeder screws. The pedal operator or a third person keeps an eye on the fluid level in the master cylinder and refills it when necessary. You'll also need some basic equipment; a flare-nut or box-end wrench to open and close the bleeders, a 12-in. or so length of clear-plastic tubing and a clear jar, preferably plastic so it won't break if dropped.

Don't forget fluid. You'll need a gallon of fresh brake fluid with a Department of Transportation (DOT) 4 rating, minimum. Once you've collected these items and a friend or two, proceed with manual-bleeding your Mustang's brake system.

Sequence is important: Start by bleeding

the wheel cylinder that's farthest from the master cylinder and work toward the one that's closest. For a single brake system (one master-cylinder reservoir) begin at the right rear brake, then the left rear, right front and, finally, the left front. For a dual system (two reservoirs) you can start at the front or rear, but do the right rear before the left rear and the right front before the left front. Remember: *The rear reservoir of a dual-system master cylinder is for the front brakes and the front is for the rears.*

First, fill the reservoir(s) with fluid while being careful not to spill any on your car's new paint. Glycol-based brake fluid could double as paint remover. As for silicone fluid, a mess is all you'll have to worry about. Some rags placed under the master cylinder should catch any spilled fluid.

Now, with all bleeders closed, fit the flare-nut or box-end wrench to the bleeder screw, then push the hose over it. With about 3 in. of clean brake fluid in the jar,

Rubber boot at fuel-tank filler tube should be replaced. A small crack can mean gasoline in trunk.

place the free end of the hose in the jar so its end is immersed in fluid.

You're ready to start: Have your "bleeding" partner *slowly* pump the brake pedal, then hold it down and keep it there—not pushing hard—while you open the bleeder. When all pressure is relieved, close the bleeder. Repeat the cycle by having your partner release the pedal and pump again.

Don't get impatient. It will take several cycles before you "get a pedal." At first nothing but air will come out of the hose. Eventually, some fluid mixed with a lot of bubbles will begin to appear. The pedal will begin to "build," but not much at first. When it seems like you can't get any more pedal, go to the next wheel cylinder in sequence and repeat the bleeding process.

Don't forget to keep an eye on fluid level in the master-cylinder reservoir. You'll be adding a lot of fluid at first to make up for what's pushed into the empty brake system. Let the fluid level drop to the bottom and you'll pump air, not fluid, into the brake system.

As you pull the hose off the bleeder, fluid will run out of the hose. When you bleed the next wheel cylinder, air will be pushed out of the hose. This is not a problem on the first go 'round because there's nothing but air in the system anyway. However, things will change once the pedal begins to build. This column of air will then be followed by what's in that branch of the hydraulic system—usually fluid with diminishing sizes and numbers of bubbles

as fluid begins to fill the system.

After you've bled all the wheel cylinders once, be prepared to do it again—and possibly again and again—until there are no bubbles *and the pedal is firm*. Don't take chances! Failed brakes could ruin your car, day *and* you.

FUEL TANK & LINES

Unless you've had your fuel tank zinc-plated, be sure all body sealer has been removed from its mounting flange and from the fuel-tank mating surface in the luggage compartment. If the fuel-gage sending unit was removed, install it now.

Lower the fuel tank into its cutout in the luggage-compartment floorpan and install the eleven mounting screws. Fit the rubber hose over the fuel-tank neck and secure with a clamp. Slip another clamp for the filler neck over the hose. Run a 1/4-in. bead of body sealer around the filler neck where the attaching flange mates to the body panel. Pass the filler neck through the rear body panel and into the hose connector. Fasten with four sheet-metal screws. Tighten the hose clamp.

Go under the car and connect the fuel-gage sending-unit wire. Connect the fuel line from the fuel pump along the transmission tunnel (and clamped to the brake line) to the fuel tank. On late-model cars so equipped, reconnect the fuel-tank vent hose.

RADIATOR

The radiator is mounted to the radiator support with four bolts. After mounting the radiator, install the overflow hose and route it down the left side of the radiator. Attach the upper and lower radiator hoses. If equipped with automatic transmission, connect the oil-cooler lines, using a flare-nut wrench. When the engine can run, fill the radiator with a mixture of water and ethylene-glycol coolant according to your area's temperature range. Follow the directions on the coolant container. Do not exceed a mixture of 50% ethylene glycol. Start the engine, run for a few minutes and then check the coolant level. Add coolant mixture as needed.

Oil Cooler—If so equipped, install the engine oil cooler now. Two rubber brackets support the cooler in front of the radiator support. The top of the cooler is retained by a single rubber-covered bracket that bolts to the top of the radiator support. Get the cooler in place, then lay the two cooler hoses between the oil-filter pad and cooler. Once everything is in place, tighten the single upper bracket bolt followed by the

hose ends.

Remember, if you have an oil cooler, the two horns are both mounted on the right side of the radiator support. If there is no cooler, the horns are mounted one on each side of the radiator support.

ENGINE COMPARTMENT

Obviously, if your Mustang was original when you took it apart, merely put it back together the same way. But if you are restoring a basket-case, or making one good Mustang from several parts cars, here's a few assembly tips.

Install the battery box against the right-front fender apron just behind the radiator support. Attach the starter relay to the right fender apron between the battery box and shock tower. The voltage regulator is fixed to the backside of the radiator support on the left, midway down. On all but early '65s, horns mount on the front side of the radiator support with the bell facing down. The early-'65 horn is mounted along the left inner-fender well.

Hang the windshield-washer bag from the bracket on the left fender apron. If so equipped, mount the windshield-washer pump below and to the rear of the washer bag. Route the hose from the washer bag to the large fitting on the pump. Run the small hose from the pump to the windshield-washer emitter mounted at the very top of the dash panel.

All electrical and mechanical components associated with the engine and chassis should now be in place. You can now route the wiring harnesses, make the connections and, if you want to, try starting the engine. If possible, I strongly recommend that any engine work that may be required be done now while the car is still without the fenders and nose. It will save some scratched paint, I'm sure.

ELECTRICAL

At first glance, a car's electrical wiring looks like a very large plate of spaghetti. Fortunately, the large plate of spaghetti can be divided into five small plates that are a little easier to digest. If you are the proud owner of a convertible, you'll be dealing with six of these. Let's call them wiring harnesses. Each harness can be approached as a system to make things even clearer.

The harnesses should have been marked during removal. If you purchased new harnesses, I'll refer to their Ford code number so you can put the right harness in the correct position.

Instrument-Panel Harness (14401)—This harness services all of the instruments,

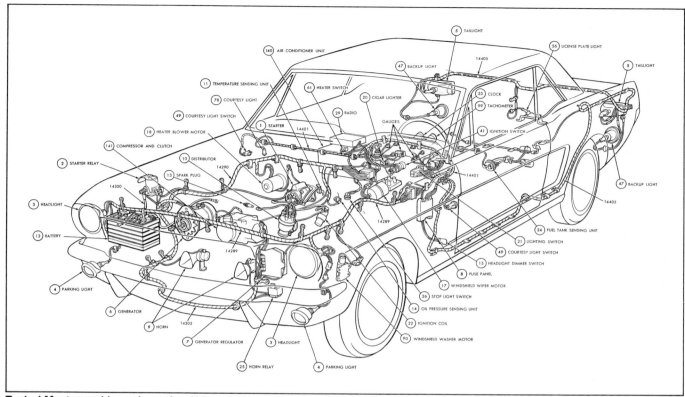

Typical Mustang wiring schematic—'65—66 shown. Courtesy Ford Motor Company.

gages, radio, ignition switch, courtesy light, windshield-wiper motor and all accessories. It also has a breakout, or dropout—a smaller wire coming out of the main harness—going to the console.

Buy a dozen large harness clamps and four-dozen small clamps to fasten the harnesses to the car body. The large clamps are for the instrument-panel harness while the small clamps will be used on the remaining small-diameter harnesses.

Suspend the instrument-panel harness inside the instrument panel. Note the two through-the-firewall connectors. These go on the left by the fuse panel. One goes to the left of the master cylinder and the other goes on the right of the brake and clutch pedals. Fix the harness to the inside of the instrument panel with seven harness clamps. Push the two firewall connectors into the firewall. On 1965—66 models, mount the fusebox to the left-side cowl panel next to the parking brake. From 1967 on, the fuse panel is mounted above the accelerator pedal on the inside of the firewall. Route the one breakout down to the dimmer switch. Feed the connectors for the courtesy lights through the cowl panel and into the door post.

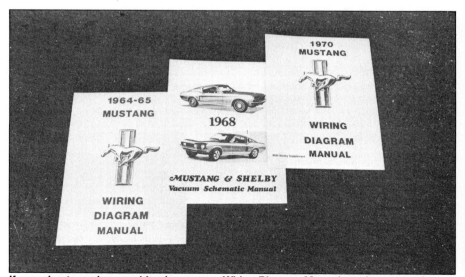

If you plan to replace a wiring harness, a Wiring Diagram Manual, such as one of these from Jim Osborn Reproductions, Lawrenceville, GA, will be an indispensable aid. Photo by Ron Sessions.

New sheet metal, such as a fender, usually comes without holes for trim fasteners. To transfer holes, make a paper template.

Align template with a vertical line, such as door opening, and horizontal line, such as rocker panel, and poke holes with a pencil. Then, tape template to new panel and pinpunch, then drill holes to size.

If you have a Rally-Pac, position the breakout for it and bring it forward along the steering column. It has three wires at the end of the loom. One is black. The other two are blue, one with a black tracer, the other with a red tracer. This locates the harness. Do not install the instruments at this time. Finish routing the other harnesses first.

If the car is equipped with an under-dash A/C unit, a single power lead (yellow tracer) runs from the A/C unit to the accessory terminal on the ignition switch. From there another wire (black tracer) passes through the firewall, terminating at the compressor clutch.

Headlight Harness (14289)—This harness begins at the left-most firewall connector from the instrument-panel harness. It runs along the top of the left fender apron and across the radiator support (under the lip at the top). A breakout is provided for the electric windshield-washer pump (on early models) and then a breakout for the left headlight and parking light. This is routed through the left fender apron at a 1-1/2-in. hole about 4-in. down from the top of the apron.

The next breakout goes to the voltage regulator. Finally, at the end of the harness are the connectors for the right headlight and turn signal. Route this through the hole in the right fender apron corresponding to the hole on the left.

Use new plastic loop-clamps to hold this wiring. There are a total of 14 clamps in the '65—66 engine compartment. The plastic connectors push into holes in the inner

fender, then loop around the wire to snap onto themselves again.

Firewall-To-Engine Harness (14290)— This harness connects to the right-side firewall connector, travels along the firewall, under the brace from the firewall to the shock tower, along the right fender apron and ends at the starter relay.

On early models, the first breakout travels to the brake-light switch on the master cylinder. The next breakout goes to the oil-pressure-sending unit and then to the coil. The next breakout has two connectors for the heater fan. The harness terminates at the starter relay.

Generator Harness (4305)—The generator harness begins at the voltage regulator with a breakout to the horn relay and to the left horn. It enters a 1-1/2-in. hole in the lower radiator support at the left side and exits the same support on the right side. Another breakout services the right horn. The harness terminates at the alternator.

Instrument-Panel-To-Rear-Lighting Harness (14405)—This harness services everything in the rear of the car. It begins at the instrument panel and fuse panel. It travels inside the left cowl panel, under the left sill plate (atop the rocker panel), over the inner fender to the left-rear taillight. Before it reaches the taillight assembly, there is a breakout to the fuel-tank sending unit. This travels along the back edge of the luggage compartment, through a hole on the right side of the luggage-compartment floor, and into the fuel tank. After connecting to the left-rear taillight assembly, the harness travels across the back panel with a

breakout for the license-plate light and ends at the right taillight assembly.

Power-Top Harness (15A668)—The power-top harness begins at the connector of the top-control switch. It routes along with the instrument-panel harness to the right cowl panel. On early-model cars, midway along the instrument panel, there is a breakout that connects to a small wire (15A669) coming from a 20-amp circuit breaker on the starter relay. This is where the top-control motor gets its power.

From the cowl panel, the harness travels under the sill plate, over the inner fender, behind the seat-back frame to the top-control motor.

Final Harness Connections—This completes installation of the six major wiring harnesses. Go back now, and with your identification tags, connect each wire to its respective component.

OK, so you didn't believe me when I said tag everything. Or, you have a car that suffered an engine-compartment fire. In either case, you'll need more help than this book can give you. Ford produced a comprehensive wiring guide for dealership use. Many Ford agencies still have these diagrams. Often, for a small service charge, they will photostat a copy of the page(s) you need. Better yet, wiring books for each year are available through Jim Osborn Reproductions, 101 Ridgecrest Dr., Lawrenceville, GA 30245. Jim has both the *Mustang Wiring Diagram* series and another for vacuum hoses.

Short of these two helps, a half-dozen jumper wires, a continuity tester and a 12-

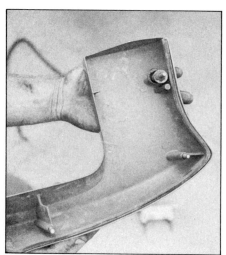

When installing rear-quarter caps, don't forget rubber seal. If omitted, you get a rattle that's hard to find.

Easy way to paint around chromed areas is to spray everything, then rub paint off brightwork with extra-fine (OOO) steel wool.

Check wiring connections carefully during assembly. Often, you'll find broken wires. Finding them now eliminates pain of tracking them down after car is assembled.

volt test lamp will have to be your sole companions to make your connections. Or perhaps you have a friend or club member with a similar car, or the local used-car lot or wrecking yard will let you take a peek at some of their iron.

EXTERIOR SHEET METAL

After all painting, rechroming, suspension and drive-line work, it's time to install those body panels taken off earlier. For most restorations, this means hanging the doors, fenders, hood and front sheet metal such as the front valance, grille and so on.

Unlike many other cars, Mustang sheet metal is easy to work with and align. A minimum of shim points and complex shapes means panels go on without a lot of bending or shimming. On the other hand, Mustangs were mass-produced cars, and the fit of some parts is not the finest. If you are after an exceptional-looking Mustang, one that fits better than a car Ford built, it will take some parts swapping and a fair amount of aligning before parts like the headlight doors and fender caps fit the way you want.

Front-end sheet metal alignment starts with the doors. The doors must go on first and align with the rear quarter panel and rocker panel. Once the gaps between the door and these two sections are equal, and the top of the door is even with the top of the rear fender, the front fender can be installed.

After both sides are done, the front sheet metal and hood can go on. This procedure allows any minor variations to be adjusted

out at the front sheet metal. If you try to hang the front fenders, hood and grille, then go and install the doors, you'll find it difficult or impossible to get the proper fit. What makes the front of the door align with the front fender will cause a misalignment at the rear, and vice versa. So always start with the doors.

DOORS

Door Hinges—Before getting the door on the car, think about the hinges. Mustang doors are large and heavy. It's a rare car that doesn't have worn out hinges, as shown by squeaky, graunchy operation and difficult closure. If the door rises slightly as its latch slides onto the striker, you know the hinge is worn out.

There are two ways to go with the hinges, either outright replacement or installing a hinge kit. Simply replacing the hinges is the easiest answer. A hinge kit uses a bushing, which you must install. So, you can spend the money on new hinges, or take the time to install the cheaper kit.

Other door parts can go on now, like the wing vent, glass and so on. The problem with installing this gear while the door is off the car is supporting the door. If the door is already painted, it is better to hang the door while gutted, then install the glass, latch and wing vent. If still in primer, then you can install the door parts now. A help supporting the door is a large inner tube from a truck or tractor tire. Inflate the tube, lay the door atop it, and presto, you've got a fixture that has enough surface friction to hold the door from sliding while you work

When bolting on front sheet metal, be sure to insert any shims that may have been removed at disassembly time. Otherwise, panels may not align. Photo by Ron Sessions.

on it, but doesn't harm the metal in any way. Just make sure the valve stem faces down against the bench or floor.

Door-Latch Assembly—Mount the exterior door handle with its two screws from inside the door. Position the lock cylinder and fix it to the door with its large, U-shaped retainer. It will take a little encouragement with a plastic mallet to seat it all the way. Position the latch in the door and mount it with its two countersunk Phillips-head screws.

There are three rods for the latch assembly: short, medium and long. If you re-

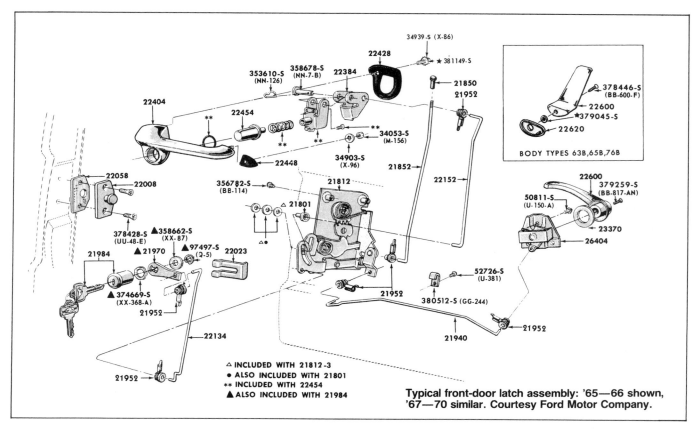

34939-S (X-86)

381149-S

22428

378446-S (BB-600-F)

22600
379045-S

22620

BODY TYPES 63B,65B,76B

353610-S (NN-126)

358678-S (NN-7-B)

22384

21850

21952

22404

22454

34053-S (M-156)

22448

34903-S (X-96)

21852

22152

22600
379259-S (BB-817-AN)

50811-S (U-150-A)

23370

26404

22058

22008

356782-S (BB-114)

21812

21801

378428-S (UU-48-E)

358662-S (XX-87)

97497-S (Q-5)

22023

21984

21970

52726-S (U-381)

21952

380512-S (GG-244)

374669-S (XX-368-A)

21952

21940

21952

22134

21952

△ INCLUDED WITH 21812-3
● ALSO INCLUDED WITH 21801
** INCLUDED WITH 22454
▲ ALSO INCLUDED WITH 21984

Typical front-door latch assembly: '65—66 shown, '67—70 similar. Courtesy Ford Motor Company.

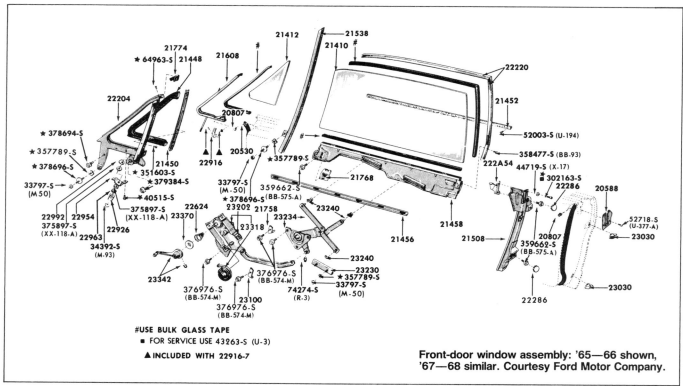

21412

21538

21774

64963-S

21448

21608

#

21410

#

22220

22204

21452

378694-S

20807

357789-S

52003-S (U-194)

378696-S

20530

357789-S

358477-S (BB-93)

33797-S (M-50)

21450

351603-S

379384-S

33797-S (M-50)

359662-S

21768

222A54

44719-S (X-17)

302163-S

22286

20588

40515-S

378696-S (BB-575-A)

21758

23240

21456

21458

21508

52718-S (U-377-A)

22992

22954

375897-S (XX-118-A)

22624

23370

23202

23234

21940

23030

22963

22926

23318

23240

20807
359662-S (BB-575-A)

34392-S (M-93)

23342

376976-S (BB-574-M)

23230

376976-S (BB-574-M)

74274-S (R-3)

357789-S

33797-S (M-50)

22286

23030

23100

376976-S (BB-574-M)

375897-S (XX-118-A)

22916

#USE BULK GLASS TAPE
■ FOR SERVICE USE 43263-S (U-3)
▲ INCLUDED WITH 22916-7

Front-door window assembly: '65—66 shown, '67—68 similar. Courtesy Ford Motor Company.

moved them from the latch assembly, here's how to get them back. The short one is the latch-control-to-cylinder rod. The middle size is the latch-actuating rod. The long one is the push-button rod.

Connect the latch-control-to-cylinder rod to the lower outboard hole in the latch assembly. The latch-actuating rod goes directly above it in the long groove. The push-button rod connects to the lock lever on the latch assembly at the inboard bottom hole. Now connect the remote-control link from the inside door handle.

Vent Window (1965—68 only)—Be sure the two adjusting setscrews are still in the frame. Insert the vent window and frame into the door. Install the two vent-window retaining screws and the setscrew locknuts.

Window Glass (1965—68)—The vent-window frame is the front run for the window glass. Leave the vent-window frame loose so there is a little play when you install the window. Position the rear guide for the window in the door but do not connect it yet.

Install two new rollers in the channel-guide bracket. The pins in the regulator arms will go in here. There is also a roller at the rear of the window that fits in the rear guide. Position the glass with its channel and new rollers into the door with the front edge of the glass in the front run and the rear rollers in the window rear guide. Loosely install the window upper-front stop and the rear upper-stop bracket. Push the pins in the regulator arms into the two new rollers. Fasten the rear guide in place.

The windows are now in. After adjusting the doors, you'll adjust the windows.

Window Glass (1969—70)—Window glass on the '69—70 models is much easier to replace than on earlier models. One reason is the absence of a vent window.

A round bar in the front and a round bar in the rear of the door comprise the front and rear runs. Bracket assemblies are bolted or glued to the window. Position the window in the door. Slide the guides onto the glass run and bolt the glass-bracket assemblies to the guides. Install the regulator arms by bolting the front regulator arm to the glass-bracket assembly and inserting the regulator-arm pin into the roller in the rear glass-bracket-assembly channel bracket. Install and adjust the front and rear stops.

Door Installation & Adjustment—Hanging Mustang doors is a time-consuming operation that must be done right. It's best to have two people for this job; one to support and align the door while the other fiddles with the hinge bolts. If you're all alone on this one, support the

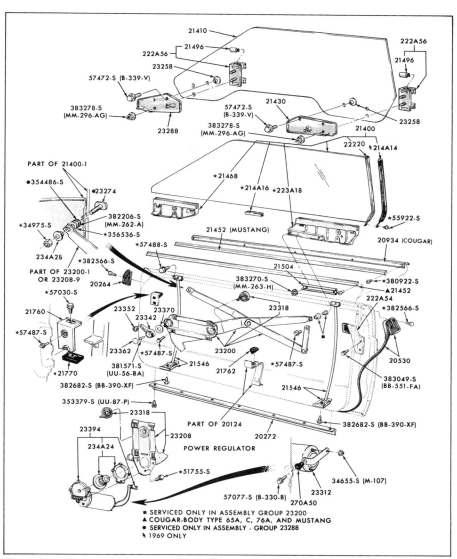

Front-door window assembly: '69—70. Courtesy Ford Motor Company.

door with a floor jack, using a 2x4 on the jack pad to cushion the bottom of the door. Drop a door now and all of that beautiful paint and bodywork may get dinged and scratched up—not to mention smashed toes and fingers!

At the body, the hinge bolts screw into weld nuts on a plate that "floats" inside the windshield pillar. To start the first hinge bolt from each hinge, it sometimes helps to use a large Phillips-head screwdriver to align the weld nuts to accept the bolts.

Adjust the door at the hinges until the door is centered in the door opening. Do this by getting the door generally in position and starting all the hinge bolts until they are snugged fairly tight. Start to close the door, watching the gap between at the

rear fender and rocker panel. Your goal is to get these gaps as even as possible along their entire length.

It will take numerous trials to get the doors correctly aligned—an hour per door is not unusual if you want a really good fit. Part of the trouble is when the bolts are loosened at the hinge, the door's weight pulls the door down, affecting your adjustment. Supporting the door at the outer end with wood, a jack stand, milk carton or whatever really helps. Also, loosen the bolts just enough so the door will move when bumped with the palm of your hand, but not so loose that the hinges slip.

Work with both gaps simultaneously. You can't get one gap set perfectly, then go for the next one. You'll get the the first one

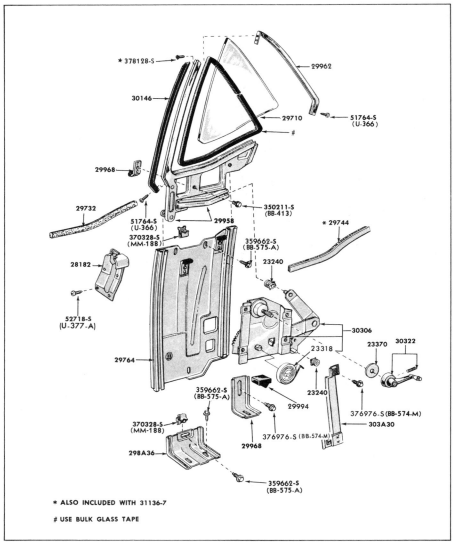

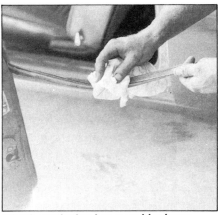

Weatherstripping has a mold release agent applied to it that can also release cement used to attach it to door. Use a wax-and-silicone remover such as R/M Pre-Kleeno to remove this agent before attempting to cement to door. *Do not use lacquer thinner because it will attack the rubber.*

Notchback and convertible rear-quarter window assembly: '65—66 shown, '67—68 similar. Courtesy Ford Motor Company.

hopelessly out of whack setting the second gap. Just take your time and constantly readjust until the doors fit correctly. Remember, the rest of the front sheet metal will use the doors as their alignment gages, so get the doors right on.

Now, the world isn't perfect, and neither are Mustang doors. When you consider both gaps, plus the body sculpture lines and door height relative to the top of the quarter panel, you will most likely have to compromise. If so, try for the most even vertical gap between quarter panel and door, plus door height relative to the quarter panel. In other words, it is better to have a slightly wider gap at the bottom so the top will be even with the quarter panel. You see a lot more of the quarter-panel height than you do the bottom gap. Plus, if the vertical gap is straight, there is less chance the door is cocked in relation to the front fenders and rear quarters.

Latch-Striker Adjustment—Only after the doors are correctly hung should you concern yourself with the latches. The latch mechanism in the door is not adjustable. All latch adjustments must be made at the latch striker. Move it laterally and vertically to get easy door closure. Do not try to overcome poor door adjustment with the latch striker. If, when adjusting the latch striker, the end of the pin connects with the latch assembly, place a heavy paper shim between the striker and door jam.

Vent-Window Adjustment—The two setscrews adjust the fit of the vent window to the window pillar. The vent-window retaining screw allows some adjustment front-to-rear. Adjust the vent windows to the windshield pillars and tighten the setscrew locknuts.

Door-Window Adjustment (1965—68)—On 1965—68 models, adjust the window rear guide for correct tilt. Tighten the two attaching bolts and the setscrew locknuts.

On 1969—70 models, adjust the window to the front windshield post and quarter window (in-and-out tilt) by adjusting the front and rear runs at the bottom underside of the door.

Now the doors and window glass are adjusted turn your attention to the quarter windows.

NOTCHBACK QUARTER WINDOWS Installation—If you removed the regulator assembly, install it first. Be sure the actuator arm is in the actuator-arm guide. Be sure the quarter-window guide-assembly lower bracket is installed. Set the quarter-window-guide assembly in place, but do not fasten it yet. Lower the quarter window into the opening, then guide the front and rear rollers into their respective runs in the guide assembly. Connect the window regulator to the roller in the glass channel. Now bolt the guide assembly to the lower guide bracket. Install the two upper retaining bolts that fix the guide assembly.

Some body trim, such as rocker-panel moldings, slide into clips riveted to body.

At either end of molding is stud fastener, retained by nut inside front and rear wheel openings.

Finally, install the two window upper stops. These go on the outside bottom of the window frame on '65—68 models. From '69 up, the front stop is on the inside lock pillar (door jam) and is retained by two screws on the outside of the pillar.

Adjustment—Roll the window up to within 2—3 in. of closure. Loosen the three guide-assembly retaining bolts and the window-regulator stop. Adjust the quarter window fore-and-aft, in-and-out to align with the roof pillar and door windows. Tighten the four bolts and the adjustment is complete.

FASTBACK QUARTER AIR VENTS

The vent assembly consists of outer metal ducting and an inner plastic shell with "doors" to control the flow of air. To assemble the vent to the car, position the outside vent in the body opening. From inside the car, fasten the vent to the framework with the six retaining nuts. While inside the car, place the plastic vent panel in place and fasten with sheet-metal screws. Finish by installing the lower garnish molding, also with sheet-metal screws.

SHELBY QUARTER WINDOWS— 1966

The '66 Shelby quarter windows are also a two-piece arrangement. Mount the plastic window in the two-piece aluminum frame with a very thin bead of sealer. Fit a new weatherstrip, slide the unit into the body

opening and attach with a 1/8-in. rivet at each end. Replace the plastic vent panel and screw it in place. Now, fit the two pieces of window molding into the panel and secure them in place with 1/8-in. rivets in each of the rivet holes.

WINDSHIELD & BACKLITE

Installation (1965—68)—To shorten the explanation, the term *window* will be taken to mean both windshield and backlite in the following paragraphs.

If you're doing a complete restoration on a notchback or fastback, do not install the windows until you have installed the headliner. The headliner glues to the window-opening flange, underneath the glass. The headliner can be replaced without removing the windows, but it makes a much neater job if the windows are out.

Begin by replacing all of the window-molding retainers. If a welded retainer stud is broken, use a 1/2-in. long #6 sheet-metal screw to hold the retainer in place.

Open the weatherstrip (window rubber) and apply a bead of sealer in the glass channel. Mount the weatherstrip to the window. Select a *strong* cord about 2-ft longer than the circumference of the window and insert this in the body-opening groove of the weatherstrip, beginning and ending at the bottom.

Now, apply another bead of sealer around the body-opening outer flange. With the help of a friend, set the window and its weatherstrip into the body opening with the cord inside the car.

While your friend pushes on the outside of the window, pull on one end of the cord, lifting the weatherstrip lip over the window-opening flange. Work from the outboad ends of the windshield in toward the top-center.

Be sure the window is well-seated into the frame. Push gently in areas where the weatherstrip lip is not fully closed over the flange.

On hardtop models, snap the window moldings into place. Use the heel of your hand, never a mallet. If you must get "physical" with the molding, lay a section of 2x4 over the molding and then attack with a mallet.

On convertibles, snap in the lower and side moldings. The top molding is retained by screws.

Installation (1969—70)—On these models, the windshield and fastback backlite are *glued in* with butyl-rubber adhesive. Only the notchback (model 65) backlite uses a weatherstrip-type seal and is installed like '65—68 models.

To install the windshield, purchase a windshield-installation kit of butyl tape and tape primer (Ford C6AZ 19562-B or equivalent). Any auto-glass shop should have the kit.

Be sure you have reinstalled all of the molding retainers as described above, because once the glass is glued in, the retainers are difficult to remove and install and you risk chipping the edge of the glass. Install spacers in the lower corners of the windshield opening. These position the

RESTORING BODY GRAPHICS

After this Boss 302 was painted and assembled, it was time to install Boss 302 graphics. First, hood is measured off and masking tape laid out to guide application. Then, hood and decals are soaked with soapy water. This keeps decal movable for fine adjustments.

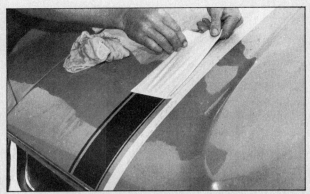

After some jockeying around to align decal, backing is pulled off.

Work air and soap bubbles to outside with rubber squeegee used to apply body filler.

Trim off excess with single-edge razor blade.

Aligning side stripes is part science, part art. Eyeball stripe from 20 ft back to check overall alignment before pulling backing. Pull backing at a shallow angle.

Getting stripes to stick around corners can be difficult. Use a heat gun or hair dryer to dry transfer sufficiently so it will tack onto edge of fender.

To install '65—68 windshield or backlite, insert a heavy cord behind lip of rubber seal. Lay window and seal against window opening and pull cord to bring seal edge around body opening.

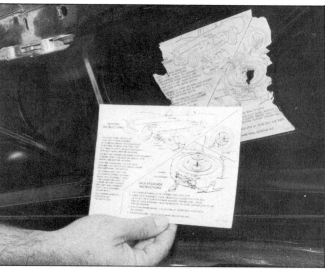

When detailing trunk area, don't forget to replace worn out, torn or missing decals. This jack-instructions decal is available from Jim Osborn Reproductions.

Before fitting new door window weatherstrip into channel, run a fat bead of weatherstrip adhesive. Allow it to get tacky before installing weatherstrip. Black adhesive is available from 3M and blends in better with black weatherstrip.

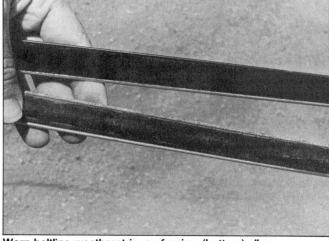

Worn beltline weatherstrip, or *fuzzies*, (bottom) allow excessive amounts of water to drain into door. Window glass must be below beltline to replace these. Removing glass from door *or* undoing window regulator to allow glass to slide down farther in door does the trick. Look for screw at door jamb edge of strip.

windshield in the center of the opening. If your's is a vinyl top, mask it off to prevent damage from the primer.

First do a dry run. Position the windshield in the windshield-opening and adjust right or left for even fit. Mark the location with a crayon and remove the glass.

Apply primer from the kit around the entire perimeter of the windshield-opening flange. Allow this to dry about 15 minutes. Apply the butyl tape to the flange. Keep it from hanging over the edge of the flange.

Be careful not to stretch the tape, particularly in the corners. Also clean the mating surfaces of the glass for a good bond.

With a friend's help, gently set the windshield in place, aligning the crayon marks. Be careful as *you only get one chance!* Adhesion is almost immediate upon contact. Now, press around the edges of the windshield until you can see that the glass

has made contact with the tape in all locations. If any areas fail to bond, apply some sealer to the edge of the glass in that location. Install the moldings as described earlier and the job is complete.

At this point, your car should be fully assembled except for the interior or convertible top. These jobs are covered in following chapters.

10

Interior & Upholstery

You don't have to be an interior decorator to enjoy simple beauty of '66 Mustang Pony Interior. This car says open the door, sit down and drive me. Photo by Ron Sessions.

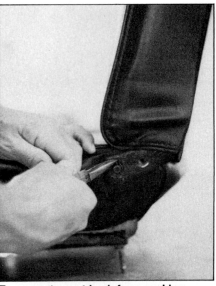

To separate seat back from cushion, remove this clip then pry seat-back leg away from pivot on cushion.

Upholstering your newly restored car puts the finish on the project. Soon, the inside of your Mustang will look like it did when it was new.

In the back of this book, along with all of the parts suppliers, you'll find those people who sell upholstery kits. These kits are reproduced from original Ford patterns. They've been cut, fit and sewn and are ready to install. With the following directions, you'll be able to install one of these kits without a hitch. Let's begin with the hardest part first, the seat covers.

SEAT UPHOLSTERY

So that we all sing from the same sheet of music, let's define some terms before you start to work—a little upholsterese, if you will. The whole seating unit is referred to as the seat—the front seat and the back seat. These are further identified as the *seat cushion* (the part you sit on) and the *seat back* (the part you lean back against). These parts can now be identified as the *front cushion*, *rear cushion*, *front back* and

rear back. The front back has one further deliniation—the *inside back* and the *outside back*. You lean against the inside back and you stare at the outside back if you are sitting in the rear seat. Got that? Good!

The part you set your feet on is called the *carpet*. A rug is something you set on top of a carpet to protect it. If you're driving the car, the vinyl mat under your feet is called the *heel pad*. The part that covers the sides of the cowl (in front of the doors) is identified as the *kick panel*. The plastic snap-on trim around the door frame is called *windlace*. The only other confusing term is what we call the panels to the sides of the rear seat. These are called *quarter panels* (or quarter trim panels), not to be confused with the exterior quarter panels. Oh yes, the material over your head is the *headliner*. Now, with this out of the way, let's tear into the front seat.

FRONT SEAT CUSHION
& SEAT BACK
Disassembly—The following operations

are similar for both bench seats and bucket seats. With the seat out of the car and on the bench, the first step is to separate the back from the cushion. Remove the aluminum side trim (1965—67) or plastic leg trim (1968—70) by removing the sheet-metal screws that retain them to the frame. Remove the spring clip from the pivot pin on each side. Pry the outside seat-back arm off the pivot pin and separate the back from the cushion. Remove the latch lever, if so equipped.

HOG-RING PLIERS

A pair of hog-ring pliers and a bar of staples are available from most upholstery-kit dealers at a very modest price. If you have a choice, pop loose a few extra dollars for a pair of long-handled hog-ring pliers. Using short-handled pliers requires a lot of strength to use them. Often, the result is blisters on your hands. For only $2—3 more, you can buy a pair of professional hog-ring pliers. I strongly suggest making this extra investment in a good pair of pliers. The money will be well spent.

Remove seat-back adjuster stops as well as all other hardware. Store in soup or coffee can.

Pry off back panel for seat back with flat-blade screwdriver.

After disconnecting return springs for adjuster, unbolt tracks from seat frame. Impact driver helps here.

Old seat covers are headed for dumpster anyway, so snip them at corners to ease removal and reduce strain on foam seat bun.

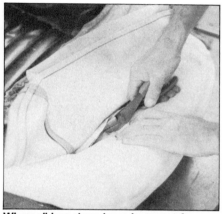

When all hog rings have been cut from outside perimeter, turn seat cover inside out and remove rings at listings holding cover down in center.

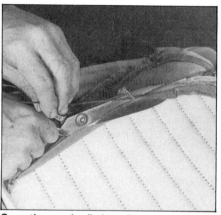

Save these wire listings from your old seat cover for reuse in new cover.

The outside back panel is retained to the back frame by the same type of clips that hold the door panel to the door. Insert a screwdriver under the clip and pry up. Be sure the screwdriver is under the clip, not beside it. If you are anywhere but directly under the clip, you'll break the composition-board back panel. Remove the stop-adjusting bolts to finish disassembling the back.

Turn your attention to the cushion. Remove the seat tracks and seat-cushion scuff plates. On bench-seat models with a center arm rest, remove the arm rest by removing the bolts that retain it to the seat. Now you are ready to remove the covers.

Remove Cover—The simplest cover of all is the bench-seat cover. Turn the cushion upside down on the bench. Note the cover is retained to the frame by a series of wire rings. These are called *hog rings* and are removed with wire cutters. Twist-off or cut the hog rings all around the seat. Remove the wires from inside the muslin strips—*listings*—and save them. They'll be used with the new cover. Turn the cushion topside up on the bench. Place one hand on the front corner of the seat holding everything down and peel the seat cover up. If you fail to hold the corner down as you pull up on the cover, you'll tear the cotton padding. Repeat this process with the other corner

and remove the seat cover. The bucket-seat cushion is only a little more complicated.

Remove the hog rings from the bottom. Save the wires for the new cover. Turn the cushion right-side up. Peel the cover up. Now, take a look under the cover and note the groove in the foam rubber. This is in a horseshoe shape through '66 and two straight grooves on most '67—70 models. The cover is retained in these grooves with a listing and wire insert. The wire and listing are fastened to the frame with hog rings. Dig down in the groove, locate and remove the hog rings. The cover will now lift off the cushion. Save the wire(s) to reuse on the new cover.

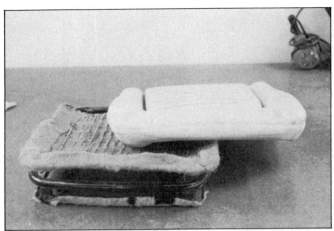

When listings have been cut away, seat cover is free and foam pad can be removed, or replaced, as necessary.

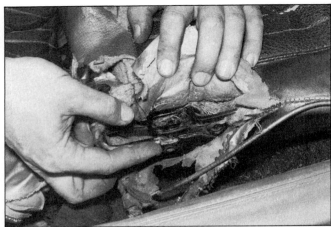

Inspect seat frame carefully. On this '68, weld broke, causing seat back to fall rearward. Repair kits are available from aftermarket.

The back covers are removed in a similar fashion. Remove any hog rings on the outside back or side of the back. Peel the cover back to access the listings. Remove the hog rings retaining the listings and the cover will come off. Now you have a naked front seat. If the padding, springs or frame has been damaged, it must be repaired before installing the new cover.

Frame & Foam Repair—Over the years, most seat covers have become damaged, exposing the internal padding. On bench seats, this is only cotton batting and may be replaced quite easily. On bucket seats, it is a different matter. A tear in the vinyl will expose the foam to the deteriorating rays of ultraviolet light. Where the polyfoam has deteriorated or been torn away, it must be replaced. Broken springs and frames must also be attended to.

Check for either of these problems. Broken frames and broken wire that supports the top edge of the springs (edge wire) must be welded. Broken zig-zag (no-sag) springs are available from auto upholstery-supply houses by the foot. If neither of these options are available to you, select a piece of welding rod the same diameter as the piece needing repair. Bend it to the shape of the broken part and wire wrap it in place with soft mechanic's wire.

If the foam is bad, most automotive-trim shops simply fill-in with cotton. This is never quite acceptable and I suggest you repair the foam by gluing in a new piece. Polyfoam, like no-sag springs, is readily available at any automotive upholstery-supply house or trim shop. Buy a piece

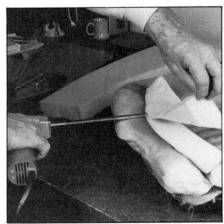

Holes in seat foam can be repaired, if necessary. After spraying seat bun and replacement foam with adhesive, whittle foam into shape with electric carving knife.

large enough to make the repair. Next, buy an aerosol can of contact adhesive specifically recommended for bonding polyfoam.

Polyfoam is easily cut with a very sharp knife, hacksaw blade or electric carving knife. And it cuts even better if you put it in the *freezer* overnight beforehand! Begin the repair by cutting away the damaged area back to strong, fresh foam. Cut a repair piece considerably larger than what was removed. Apply one *light* coat of spray adhesive to each mating surface. The carrier in the adhesive will attack the foam if the application is too heavy.

Allow the first coat to dry for five minutes and apply a second medium coat. Allow three to five minutes for this to dry and bring the two mating surfaces together. *These surfaces must mate correctly the first time. Once together, the bonding surfaces will not separate without tearing.* That's why it's called *contact* cement.

You made the repair piece considerably larger than the surrounding area. Now you must carefully trim the excess to match the original contour. Any very sharp instrument can be used for this operation. Besides those mentioned previously, single-edge razor blades, grandad's straight razor or any very sharp knife will work well. Don't use the hacksaw blade to shape the foam. It leaves a ragged edge.

Grab an edge of the foam and begin slicing off long, thin strips of the foam, working to the desired contours. This is done much like a fine chef slicing roast beef. Cut a little at a time, don't rush and keep the edges as smooth as possible. The finished product will not be as smooth as the original molded surfaces. This is acceptable because the tightly wrapped vinyl seat-cover material will flatten the cut marks and you'll not see the repair. When the repairs are completed, place the foam back on the seat frame and secure with a few hog rings.

Several companies offer reproduction polyfoam replacement seat pads. At this writing, however, I recommend sticking with original-equipment foam, if available, from a donor car. Most reproduction foam I've seen is denser and harder than the

Rear (or front) bench seat should have layer of cotton over old surface for better fit. Cut through cotton where listings will connect.

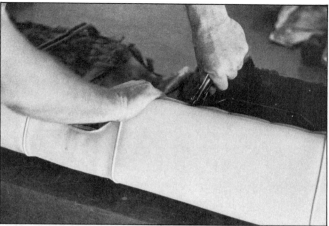

After connecting listings, hold corner true with one hand and pull cover down with other. Then, turn cushion over and ring it in place. There is no need to guess on any of the places to ring. The material fits only one way.

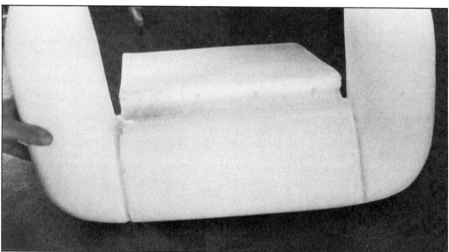

If you elect to use a reproduction seat pad, or bun, modifications may be necessary so listings connect to seat frame. On most I've seen, the transverse grooves are not cut completely through. With a sharp knife, cut transverse grooves (arrows) to within 1/2 in. of back side.

originals. The indent for the wire and listing is not deep enough to allow ringing the listing wire to the seat frame. If you still want to use reproduction foam, cut through the pad in the listing groove so that you can hog-ring through it as shown in the photo, top right.

Install Bench-Seat Cushion Cover— Again, the bench-seat cushion cover is the easiest to install. You'll need a few yards of cotton batting. This may generally be purchased by the yard or pound at any automotive-trim supply house or trim shop.

Two yards for the front bench-seat cushion and four for the back seat will do the job.

If there has been any damage to the original cotton batting, replace the damaged area with new batting. When the shape looks right, spread a layer of the new cotton directly over the old. Along the side edges, trim the new cotton flush with the old. Trim it flush along the back edge, but let it hang over the front edge. Now you're ready to install the cover.

On early bench cushions with no listings, turn the new cover wrong-side out

and set it on top of the seat. Carefully align the cover so it is centered on the cushion with no excessive differences on the left or right sides. In the same manner as you removed the cover, place one hand on a front corner and *skin* the cover down at that edge with the other hand. This will keep from pulling the cover out of alignment with the frame.

Repeat this operation at all four corners. Now the cover is on but must be hog-ringed down. Turn the cushion over, insert the wires you saved and begin ringing the cushion on. The manufacturer of the cover generally folds the cover over making a hem. This is where the wire is inserted. If the manufacturer didn't add this hem, double the material under about 2 in. and ring through the two layers of material.

Begin at the center-front. Pull the cover tight until the wire lays over the hog-ring hole, or edge wire, from where the old rings were removed. If there is no welt, pull the cover tight until there is only about 7 in. of front band (measured from the bottom of the frame to the radius of the front crown) and insert a hog ring with hog-ring pliers. Place a ring in every hole, or about every 6 in. from center left, then from center right.

Turn the seat around, pull the cover *very* tight along the rear and ring it in place, again from the center out. Now do the sides and corners and the cover is installed. For a description of the back cover, see the section below on bucket seats.

Install Bucket-Seat Cushion Cover— Bucket seats are a bit more involved because of their center indent that gives the

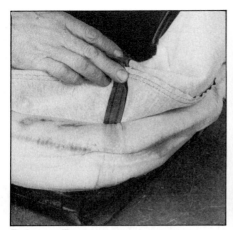

Before pulling seat cover over corners, line up seams with old marks on seat bun. With high-quality kits like ours from California Mustang Parts and Accessories, getting correct fit is easy. Once centered, hog-ring fabric to seat.

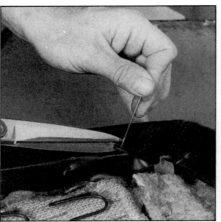

Use a pick to find holes, and scissors to make necessary openings. These are for seat-back cover.

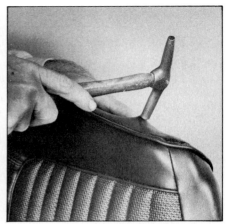

Tapping seat cushion stud with hammer will cut upholstery, making a neat, fast fit around stud.

bucket shape. Turn the cushion cover inside out and install the wire you saved from the old listing into the new listing. Lay the seat cover on top of the seat as described above, but be very careful to align it so there is an even amount of foam protruding from each side. The foam is about 1-in. larger than the cover all around.

Without moving the location of the cover, reach under it and push the wire and listing into its groove. After this, check to make sure the cover is still located correctly on top of the seat and the listing is in the groove. This care in centering prevents wrinkles and puckers; so be certain you're well-centered.

Put a hog ring in your pliers, bend the front edge of the seat cover back and ring the listing and wire into the groove; again in the center. Check once more on cover location and then finish ringing the listing in place.

To pull the cover down over the front edge of the bucket seat requires a little finesse. The top of the seat is larger than the bottom. To make things a bit easier, heat the vinyl, page 142, with a heat gun or hair drier. Warm the vinyl to a point where it is almost too hot to touch. Push the rubber back as far as possible and then skin the cover down over one corner. Reheat the other corner and pull it over. If you were very careful, especially at the seams, you'll not tear the cover. If you get in a hurry or neglect the heat, you stand a very good chance of tearing it.

If it seems that the cover is going to tear, stop. Apply a little more heat. If this doesn't work, remove the last hog ring at the rear of the listing. This will let the back of the cover come forward, giving you a little more room in front. When the cover has been pulled down properly, turn the seat over and hog-ring it down following the directions given above. Before finishing off the back edge, however, mark the holes for the scuff-plate screws; they're very hard to find after the cover is ringed down. The bucket-seat cushion is done. Now turn your attention to the bucket-seat back.

Install Bucket-Seat Back Cover—Back-cover installation is similar to cushion-cover installation. Feed the wire into the listing on the inside of the cover. Don't be afraid to bend the wire a little. This will help it slide in. Once in place, it can be bent back to the correct shape.

Once more, center the cover inside out, over the back. Push the listing and wire into the groove and hog-ring the center. Check for fit. Now, finish the hog-ringing.

Some heat on the edges of the cover will make it easier to pull around the foam. When the cover has been pulled over the foam (one corner at a time) you can begin ringing it on.

Generally, the back cover has not been hemmed for wire inserts. Therefore, you must make a judgement call on how tight to pull the cover. The top band should be about 4-in. wide after the cover has been

pulled tight. This can vary about a 1/2 in. You must, however, maintain the same width along the top band all the way across.

Pull the material tight in the top-center. Wrap the material around the hog-ring bar and set a ring. Work from the center both ways in this fashion, maintaining the curve of the back. When the top is set, turn your attention to the bottom of the back. The two bottom corners must now be pulled tight. Be sure the foam is pushed into these corners tightly with the edge wire directly under the corner of the foam. Pull the side band at the corner around and ring it to the closest no-sag spring. Repeat this with the other corner. Now, check for wrinkles along the front edges. The work you just did should have removed them. If not, check to be sure the cover was pulled tightly into the corners. Cut the rings and pull things tighter if any wrinkles remain.

Finish ringing the sides of the seat cover to the no-sag springs. You are ringing to the springs even though there appears to be a ring bar along the edge. This is not a ring bar. This is where the outside back panel fastens into the seat back. Don't put rings in these holes.

Leftover Wrinkles—When you've done the best you can and pulled the covers as tight as possible (without collapsing the seat), you may still have a wrinkle or pucker left; especially in the corners by the pleats. Small wrinkles and puckers can be removed with heat.

Again, use a heat gun if you can or the

To prevent cutting or marring cushion vinyl with pivot leg of seat back, insert a large piece of cardboard over pin.

It's easy to forget this plastic spacer on seat reveal molding. If it is omitted during installation, it will cause a concave dimple on surface of molding.

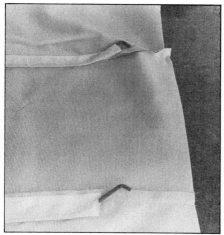

If you forgot to mark headliner bows, line them up smallest to front, larger to rear. Then, insert bows with headliner on bench.

family hair drier. Begin heating the wrinkled area for approximately 6 in. in all directions. Heat it slowly so the heat penetrates the vinyl. Continue increasing the heat to the point where the vinyl becomes too hot to touch, then just a bit more. It would be wise to experiment with this heat level on the old seat cover. Test to see how hot you can get it before the vinyl melts or burns. This way, you won't burn your new cover.

When you have reached that optimum heat around the wrinkle, stop heating the area and let it cool. As it cools, the wrinkle should disappear. You've actually shrunk the vinyl with high heat.

Install Outside-Back Panel—To finish the seat back, you must install the outside-back panel. Remove the retainer clips from the old panel and install them on the new. Replace any bent or broken clips. Set the panel aside and locate the holes in the seat back to which the panel will fasten. You can feel these holes through the vinyl. Cut the vinyl away around the holes and then set the panel onto the seat back. Align the clips in the holes. Then, with your hand doubled into a fist, give the back a sharp blow directly over the clip, driving it into the hole. Do this all the way around. Align the clip, then drive it home.

Assembly—Assemble the back to the cushion after installing the seat tracks. When putting the back onto the cushion, insert a large piece of stiff cardboard over the outside pin on the cushion. This will let

you pry on the back's pivot leg without tearing the vinyl on the cushion. The photo on this page shows cardboard and a screwdriver being used. After the leg is on the pin, tear off the cardboard.

REAR CUSHION & SEATBACK

Notchback & Convertible—The rear cushion and back on notchbacks and convertibles are upholstered in the same manner as a standard front bench seat. The cover is removed. Any wires are saved and transferred to the new seat. A layer of cotton is placed over the cushion or back, over which an inside-out cover is centered. Strip down first one front corner then the other. Repeat this at the back. Insert the wires or double the material over and hog-ring in place.

Some later-model rear bench seats have a listing on each side of the center rise that accommodates the drive-shaft tunnel. If your car has this type of seat, center the cover, hog-ring the listings to the frame, then pull the cover over the padding.

The rear back is covered in the same manner as the seat.

Fastback (2 + 2)—The fastback rear bench seat and back are handled differently. The bench seat is more like two buckets attached to one another. Cut the cover loose and pull it up. From 1966 on, there are two slits for the seat belt to pass through on each side of the center hump. Remove the rings from these and pull them through the seat. Remove the hog rings at

the listing on each side of the center hump and the cover will be free.

Reverse this process to put the new cover on. Again, turn the cover inside out and center it on the cushion. Ring the listings in place. Push the slits for the seat belts through the foam and ring these in place on the back side. Now pull the cover around the foam and frame, securing it to the frame with hog rings.

The rear back is simplicity itself. Remove the hog rings. Remove the cover and replace it with the new one. There is virtually no fitting or adjusting to do.

HEADLINER

The headliner, if done correctly, should come out looking like a factory job with no wrinkles anywhere. Or, you can do it the "Fast Freddie" way and still get a reasonably good job. I'll talk you through both ways and you decide which is best.

Factory Installation—At the factory, the headliner was put in before either the windshield or backlite was installed. Also, none of the door weatherstripping was in. To do the job right, your car should be in this condition. There should be no interior trim in the car at all. In the factory, they worked from the headliner down. You should do the same.

When you removed the old headliner, you numbered the bows from the front to the rear. Replace the bows into the headliner listings (just like the seats) in the correct order. Number one is to the front. The

SEAT VARIATIONS

Pony Interiors—If your Mustang came equipped with a standard interior, you cannot simply install a Pony Interior. The polyfoam and subframe are different. If you want a Pony Interior in your car, you'll have to buy Pony seats.

Covering the Pony seats is the same as covering a bucket seat. However, the inside listings are different. The cushion incorporates two straight listings, one on each side of the cushion. Center the cover as described above. Push the listing and wire into the groove. Ring the listing into the groove starting from the center (of the listing) and working both ways. Repeat this with the other listing and finish installing the cover as on the standard bucket.

The back incorporates three listings and is somewhat more complex. Begin with the listing at the Pony logo in the center of the seat. Insert this wire but leave the two outboard wires out for the present. Center the cover and insert the listing in its groove. Secure with two or three hog rings. Now, insert the wires in the side listings and secure them in their grooves. Finish as you would for a standard bucket seat.

1969—70 High-Back Bucket Seats—The 1969—70 models were available with a high-back bucket-seat option. Recovering these seat backs requires a bit more work. To remove the back cover, first remove the back panel by popping out the trim clips as for earlier models. Remove the latch handle. Now you have access to the cover. Remove the hog rings from all four sides. On Grande covers, remove the buttons from their spring nuts.

At the bottom sides of the back, roll the cover back until you can access the listings going down the sides of the back. Cut the hog rings loose, working from the bottom of the back to the top. It may help to alternate sides. When the listings are free, remove the wires.

Fold the cover back and remove the hog rings from the listing that retains the bottom of the head bolster. Now the cover will come off the back.

Installation is a little different, also. You must start with the head bolster. *Without* turning the cover inside out, insert a wire into the bolster listing. Pull the cover down over the seat and work the top (bolster) down tightly against the foam and frame. Fold the cover back over itself until you can reach the bolster listing and ring it in place. Turn the back over and ring the outside-back bolster listing in place.

Return to the face of the back, insert the remaining two wires into their respective listings. Working from the top down and alternating sides, ring the listings in place. The inside, or face, of the seat back is done. Finish the outside back as previously described.

Bench Seats with Listings—If your later-model bench seat incorporates a listing to indent the cushion, follow the directions above. Remember to work from the listings closest to the center first, then any along the front and finally, at the sides.

Headrest—To remove the headrest, pull up until the post exits the seat back. If it does not pull out, slide a 12-in. long piece of sheet metal or steel rule down in front of the post, pushing the spring clip out of the detent. Then remove the headrest. The guide sleeve for the post is removed after the back panel has been removed. Trip the locking tab and then remove the guide sleeve.

To fasten the new cover to the frame, purchase some #6 tacks at the local hardware store (you can purchase fewer tacks at the hardware store than at the auto-trim supply house). Pull the cover on and put a few tacks in half way to be sure the cover is straight. Make any adjustments then finish tacking the cover in place.

Center Armrest—There are two styles of center armrests. One is incorporated into the bench seat while the other is part of the center console. To recover the console model, remove the bottom panel and the staples that retain the cover. Replace with the new cover. Be sure the corners are aligned. Pull the material tight and temporarily tack each corner. Begin at the middle of one side and place a temporary tack. Continue placing tacks from the center out. Repeat on all sides. Now, check to be sure the band is the same height all the way around and drive down the tacks. Replace the bottom panel.

The bench-seat armrest is like a large sock. Remove the staples and then the cover. Heat the new cover a bit and work it onto the armrest. If it refuses to go, wrap the foam with clear plastic film. A garment bag from the dry cleaners is perfect. Now slide the cover on and leave the plastic inside. Tack the new cover on at the mounting board and the job is complete.

This covers the upholstery section of the seats. Set them aside now while you finish the rest of the trim.

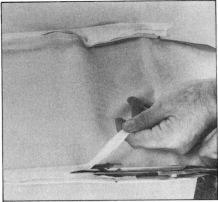

Cut back listings only enough to expose ends of bows. If you cut back on listings too far, headliner will fit poorly.

hole. Place each bow, working from front to rear. It may help to temporarily clamp the headliner to the windshield opening. At the last bow, there are two holes, called headliner-support retainers. Be sure these are secured in front of the backlite opening. Do not hook these to the rear bow at this time.

Find the penciled center line of the headliner and align this with the centers of the windshield and backlite openings. Temporarily clamp the center line in place. Move to bow #2. When the bow is located in its hole, carefully cut a slit in the listing about 1—2-in. back. Repeat this on the other side of the bow. Now stretch the headliner from one side, then the other. The idea is to remove excess material bunched up at the end of each bow and remove wrinkles along the seam.

Cut back a little more on the listing and pull out the slack. Repeat this process on bow #2 until there are no wrinkles along the stitching. The other bow stitchings will retain their wrinkles. Now, move to bow #1 and repeat the process until the seam is smooth. Repeat this for bows #3 and 4. Bow #4 will require a great deal of cutting at the listing, as much as 10 in. or more. The headliner should now be smooth but not tight.

Return to bows #2 and 3. Cut back enough listing to pull the headliner tight here. Be careful to cut only the amount needed to make a smooth line. If you cut too much, it will cause a bump to appear in that area when the headliner is complete. Using 3M Super Weatherstrip Adhesive #08001 or similar contact cement, apply a light coat to the headliner seam at bows #2

headliner will be too wide for the bows. Don't cut anything yet. Slide the headliner up onto the bow until enough bow end protrudes to place it in its respective bow

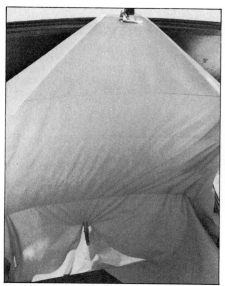

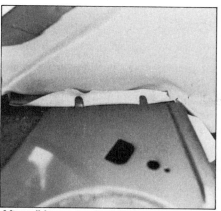

After all bows are installed, wrap headliner around cardboard strips at rear package shelf and anchor it with two metal tabs.

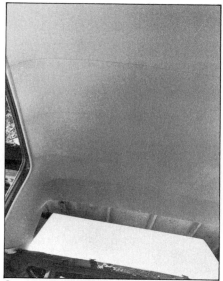

Install #1 bow first. If it helps, clamp ends of headliner in place before proceeding with rest of bows.

Completed headliner with windlace installed. Now, windshield and backlite may be replaced, followed by sunvisors, coat hooks and so on.

and 3 and to the outside lip of the door opening. Pull tightly on the seam at bow #2 and press the cemented surface to the cemented door lip. Do the same with bow #3.

Cross to the other side of bow #2 and 3, pull the fabric tight and cement that end of the two seams down. Repeat the process at bow #1. Bow #4 requires a different approach. (On fastbacks there is no #4 bow.) After bow #3, the headliner is cemented to the backlite-opening lip.

Directly below bow #4, and behind the quarter window, is a strip with protruding metal prongs. Pull the headliner taut and press over these prongs to retain.

Hook the two headliner-support retainers through the listing and over the #4 bow. When you pull the headliner tight, front to rear at the windshield opening, these retainers will prevent the bows from falling forward. When you have completed this operation, the headliner is located and you can turn your attention to cementing it in place.

Begin at the center of the windshield opening outside the car. Remove the clamp and begin cementing the headliner to the windshield opening lip, working from the center both ways. Be sure the center line of the headliner is located at the center of the windshield opening. After cementing, trim off all excess material beyond the width of the lip.

Repeat this process at the backlite opening. Things will go smoother if you clip the headliner material at the radius of the two top corners. Do not cut too deep. Avoid cutting beyond the edge of the lip.

In front of the backlite opening at each end of the package tray are tacking strips. When the headliner has been glued in at the backlite, tack it down to these strips with #1-1/2 or #2 tacks.

Finish the installation by cementing the headliner all around the door and quarter-window openings. Trim all excess.

Quick Installation—Earlier I said I would demonstrate the "Fast Freddie" method of installation. This does not require the removal of the windshield, backlite or door-opening weatherstrip. However, the windlace must be removed.

Install the headliner in place as described previously and cut back the listings until it hangs free with no bunching. Place a daub of contact cement on the front center line and in the center of the windshield header. Pull tight and contact the two surfaces. Repeat this at the backlite. This takes the place of clamping the headliner in place.

Select a new piece of windlace about 2-ft longer than the opening around the door and quarter window. Instead of pulling on the seam and cementing it to the outside of the door lip, pull the seam tight and secure it in place by snapping the windlace over it. Do this at the other side. Only secure bow

#2 at this time.

Return to the windshield area. You're going to cement the headliner to the windshield header *under* the windshield weatherstrip. Carefully trim away all but about 1 in. of headliner. Cement the headliner along the windshield header about 12 in. from the center to the right. Now, return to the door opening and pull the headliner tight and secure it with the windlace for a distance of about 12 in. Go back to the window and repeat the above process, then go back to the door opening. You'll finish at the windshield post.

Repeat the operation at the left-front, then right-rear, ending with the left-rear. Trim off excess material. You have now cemented the headliner under the windshield and backlite weatherstripping and retained it at the sides with the windlace. This is a very fast and good solution to installing the headliner without removing the windshield or backlite.

If you installed the headliner with the glass removed, the windshield and backlite may now be installed as described, page 137. Finish by installing the new windlace. If you can't snap it on with the heel of your hand, use a leather mallet to *gently* encourage it. Install the coat-hanger hooks, rear-view mirror (pre-'68) and the little metal piece on the windshield pillar called the side-front retainer.

The sun visors are all that remain to

147

Be sure to bolt on outboard rear seat belts before installing quarter trim panels.

Ice pick is great help in finding screw holes for interior trim.

Carpeting is easiest interior item to install. Biggest challenge is centering carpet and making cutouts for shifter, seat belts and so on. Small sections at bottom of fastback quarter panels are retained with adhesive.

finish the headliner area. New sun visors must be made at an upholstery shop equipped with an "industrial strength" sewing machine. So plan on farming out your visors.

DASH PAD

Regardless of year-to-year styling changes, all dash pads are held in place with bolts that are mounted to the inside of the pad. If, as advised, you did not install the instruments or other accessories under the instrument panel, there will be no problem with the dash pad. Set it in place and be sure each of the mounting bolts enters its correct hole. Thread the nuts to the bolts and the pad is secured. Now you can install the instruments and accessories. Refer back to page 45 for disassembly and pages 155 and 158 for assembly instructions.

CARPET

The carpet and pad are fairly straightforward. When purchased as a kit, the pad is glued to the molded carpet and the two are installed together as a unit. If they fit well—and they should because the carpet is molded into the floorpan shape—there is no need to cement them in. Begin the installation with the front.

Notchback Without Console—Most replacement carpet comes with a cutout for the shift lever. If your's has not been cut, lay the carpet down on the front floorpan, centering it in *front* of the shift lever. Carpet will be sticking up everywhere around the shift lever. When you're sure the carpet is centered correctly from the shift lever to the dash panel, mark and cut a hole for the shift lever. Lift the carpet up and pull it down over the shift lever, smoothing it out.

When you're sure the carpet is correctly located, cut out for the dimmer switch and A/C drain hose. Locate and cut out for the front-seat-track bolt holes.

Using some 3M Weatherstrip Adhesive, or equivalent, cement the outboard edges in place to keep the carpet from moving.
All With Console—The console must be out of the car when the front carpet is installed. Install as above after removing the console. Replace the console in the car. Use an awl to locate the screw holes in the floorpan. Push the awl into each screw hole in the console, through the carpet then wiggle the awl around till you find its hole in the sheet metal. Carefully remove the awl and insert each screw.

If you are unable to find a hole, *do not drill a new one!* The drill bit will grab a carpet thread, wind it up on the bit and give you a terrible run in the carpet. Use an awl that can be driven with a hammer to punch a new screw hole in the floorpan.
Rear Carpet—Like the front, the rear carpet just lays in there. The only critical item is cutting the seat-belt-bolt mounting holes. Lay the carpet in and center it. Lift up one side, locate the seat-belt-bolt mounting hole in the drive-shaft tunnel. With your awl, punch a hole through the carpet and into the bolt hole. Leave the awl in place and repeat this operation with another awl on the opposite side. Now cut out a hole large enough to accommodate the seat-belt-bolt and thread a bolt into the hole to keep things in place.

Finish laying the carpet out to the side, find the outboard seat-belt-bolt hole and cut out for it. Do the same for the other seat-belt-bolt holes. Finally, cement the carpet down on the outboard edges.

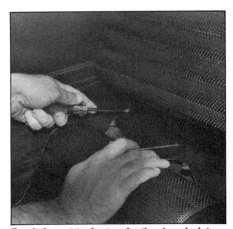

Don't forget to fasten fastback upholstery "tail" under fold-down rear seat. Pull vinyl tightly to remove any rolls or wrinkles, then punch hole and insert screw.

Fastback—Fastbacks add carpeting to the outside back of the rear seat and the luggage space behind the seat. This comes in three sections. To install the carpet on the outside back, remove the trim pieces, remove the old carpet and cement the new carpet to the boards. Trim any carpet that extends over the edge and replace the trim.

With the carpet installed, there remains only the door panels and quarter panels to install to finish the car's interior.

KICK PANELS & QUARTER PANELS

There remains only the trim panels to

install to finish the job. The kick panels are all retained by two or three sheet-metal screws into the cowl panel. Quarter panels, both steel and plastic, are also retained with sheet-metal screws. In 1969—70, the door-panel clips were incorporated along the front edge of the quarter panel. Install these quarter panels the same way you installed the door panels. Align the clips with the holes then press them in. Finish by installing the sheet-metal screws.

Quarter panels on convertibles are an exception. The top half of the panel is stamped metal, while the bottom part is metal, but upholstered in vinyl. The two parts separate with screws behind the panel. Remove the old cover, install the new and cement its edges down with contact cement.

DOOR PANELS

Through 1968, Mustang door panels were of the insert variety. That is, they were smaller than the full door. In 1969 and '70, a full door panel was used. The standard-interior door panel incorporated a bolt-on armrest while the luxury or Pony Interior had a molded armrest in the panel insert.

On door-panel inserts, install new door-panel clips. Align the panel with the indent in the door. Adjust each panel clip to fit in the hole. Now, beginning at the bottom of the panel, push each clip into the door working your way across the bottom, up the sides, then across the top.

The full door panel installs a little differently. Hook the top of the door panel over the top-inside edge of the door. Be sure the door-lock plunger extends through the hole in the panel. Adjust each clip to align with its respective hole. Push the clips into the holes working down the sides and across the bottom. Mount the armrests, door handles, window cranks, and the job is complete.

SEATS

Rear Seat—Install the rear seat before the fronts. The seat back goes in first. For a notchback or convertible, fit the back to the hooks and snap it into place. If originally used, secure the seat back at the bottom with sheet-metal screws.

For a fastback with a folding seat back, secure the back to the floor pan on each side at the pivots. With the seat back in its upright position, check the stops against the rear floor section and adjust each as required. Check also the seat-back latch at

When installing fastback folding-seat chrome trim, use a wide-blade putty knife to push carpet under trim. Be sure to get grain of carpet going in right direction before gluing it down.

Finding seat holes is easy. Drive an ice pick through carpet from bottom up, then cut hole to match bolt or stud diameter. Don't try to force seat studs through pick holes; you'll pull and misalign carpet.

It seems that replacement parts never are available in the color you want. Most are supplied in black vinyl and will have to be dyed. Armrests have a mold-release agent on them that must be removed with a mild enamel paint thinner prior to dying.

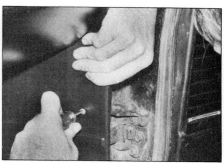

Kick panels are simple to install, but you'll have to work carpet into corner for panels to fit flush at bottom. Some downward pressure may be required to install screws, depending on thickness of new carpet.

the top right. The back should be preloaded in the latched position so it won't flop back and forth and rattle. If your fastback has a fixed-back, bolt it in place to the floor at each side and at the top right where the latch would otherwise be.

Now for the cushion: Lay it in position on the floor with the rear section immediately below the seat back, then force the cushion back with your knee to snap it into place.

Front Seats—If you haven't bolted the seat tracks to the bottom of the front seats—*seat* if you have one of the rare bench-seat models—do so now. Install the seat-track tie rod and adjust the turnbuckle so both latches release when the release handle is operated. Check also that the latch on the side opposite the handle locks when the handle is released.

Starting with either seat—driver's or passenger's—set it in place in the car so the seat-track is flat against the floor and the mounting studs project down through the holes in the floorpan. While holding the front of the seat down so it doesn't tilt backward, reach underneath the car with the other hand and thread a nut on the outboard front stud. You can let go now and thread on the remaining three nuts. Tighten them. Don't forget to install the body plugs while you're under there. Check again that the seat-track release mechanism works correctly. If necessary, readjust the tie-rod turnbuckle until it's right.

Install the opposite seat using the same methods and you're finished with the interior.

11

Instrument Panel

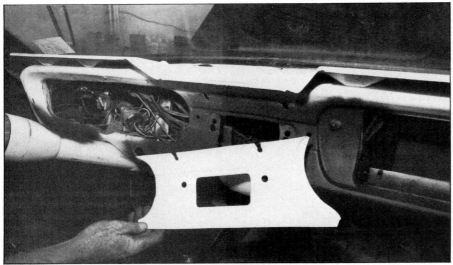

If instrument panel was hacked open to install aftermarket radio, replacement plastic cover with the correct radio cutout can be installed. Photo by Ron Sessions.

Usually, plastic instrument bezels need to be replaced on a 20-year-old car. These were scratched badly.

No matter if you are giving your Mustang a quick freshen up or the full-blown best-of-show treatment, the instrument panel is one area you want to look its best. After all, what part of the car do you look at more?

Fortunately, getting the typical Mustang dash back in shape is not difficult. Except for some of the later cars loaded with accessories, Mustang instrument panels are very easy to work on. Lots of elbow room and plenty of available replacement parts make for a speedy turnaround.

At this point, your instrument panel should be completely disassembled as outlined in Chapter 3. If not completely apart, it should be torn down as far as necessary. On panels hacked open for tape decks, everything has to go. The panel must be completely bare. Then the hole can be filled with a metal plate welded in place. Welding means grinding, filing, adding filler, sanding, prepping and finally painting—not the sort of thing to do while dodging the dash pad.

If your panel is merely tired from sitting in the sun, however, a complete teardown is probably not necessary nor even desirable. Completely tearing down a daily driver's instrument panel that's working properly and only needs a dash pad is going too far. Heater controls, ducting and so on can probably stay put, along with factory A/C. The instrument cluster and dashpad will need attention, however.

INSTRUMENT CLUSTER

Cleaning and a little touch-up paint work put most instrument clusters back in business. The *camera-case* matte finish and chrome on the bezel cannot be restored with masking and spray painting, so don't try. Just pay the price and get a new one. You could paint the black sections with a spray can, but it wouldn't come out right, and there's nothing you can do with the chrome—short of sending it out to Mr. G's Mustang City or a similar outfit for rechroming.

So, plan on a new instrument bezel. To disassemble, clean and reassemble your instrument cluster, read on.

1965—66 Models—While the rectangular '65 cluster and the five-dial '66 example differ somewhat visually, they share common design features. These clusters are individually wired. Thus, all wiring must be removed and installed connection by connection when pulling and installing these clusters from the instrument panel. See the nearby drawing if you need help rewiring your cluster. This also means any wiring damage must be handled inside the instrument panel. If the wiring is bad, a new instrument-panel harness is needed.

Once the cluster is out and on the bench, flip it over and remove the small Phillips head screws around its perimeter. There are six screws on the '65 cluster and eight on the '66. Now the instrument housing will lift up and off. Don't worry about the bent metal tabs. The housing will lift out without disturbing them.

Separating the instrument housing from the bezel frees the lens and a black-painted retaining panel. Be extra careful the lens and black panel don't slide across the instrument faces. Instrument needles are in-

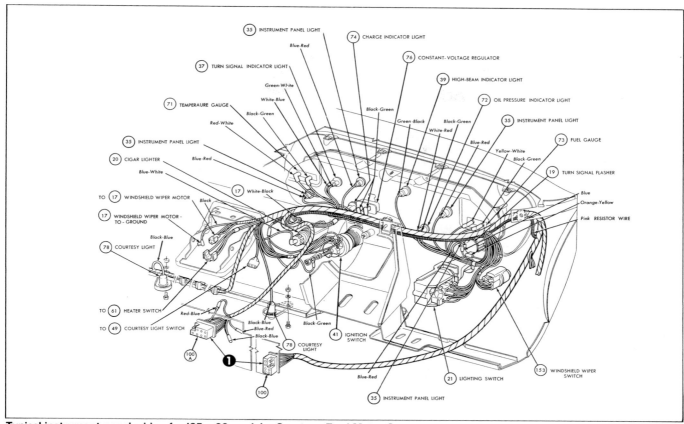

Typical instrument-panel wiring for '65—66 models. Courtesy Ford Motor Company.

credibly fragile. One slight brush, even light finger pressure, will bend and destroy the needles.

For cleaning, there is no need to remove the instruments from the housing. But if an instrument needs to be replaced, removing and replacing them is nothing more complicated than removing the two fasteners behind the instrument on the housing.

Assembly is simple, also. Lay the bezel face-down on the bench so the backside is facing you. Then, lay on the lens. It is drilled to accept the pins and screw holes only one way, so there is no need to worry about installing the lens backward. Now you can lay the metal intermediate piece in place. Use the notch and tang by the speedometer for orientation. Set the two cardboard light hoods in position on the intermediate piece. Some clear tape on their outer edges will help keep them in place while you fiddle with the assembly, but the small upturned tabs do a good job of holding them steady. Make sure the small rubber pieces that keep the lens from rubbing against the bezel are in place around

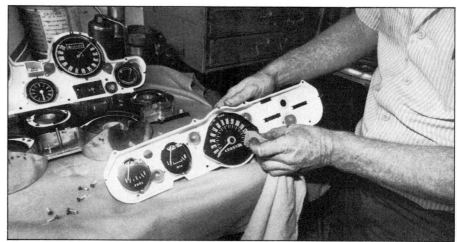

Clean instrument faces with a damp cotton shop cloth or camera cleaning brush. Use a light touch. Be careful not to bend or break the fragile needles. Photo by Ron Sessions.

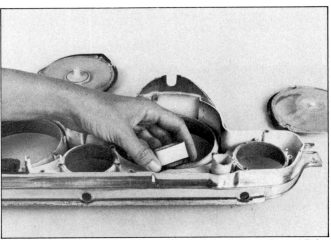

When reassembling, don't forget to install small cardboard light hoods.

If your '67—68 Mustang has a clock, first unscrew clock knob.

the instruments. The rubber stand-offs fit on upturned tabs located around the instruments.

Now you can set the instrument housing with the instruments and all related hardware onto the assembly. Install the perimeter screws and the cluster is ready to go.

1967—68 Models—Besides a completely different layout, the '67—68 cluster features its own wiring harness. So, instead of many single disconnects, these clusters have but two multi-pin connectors between the chassis and cluster wiring. This makes removal easy, but does leave some wiring on the back of the cluster.

The wiring doesn't cause any problems when it comes to servicing the cluster, however. With the cluster face-down, search its perimeter for nine Phillips head screws and remove them. Lift the housing up and out, taking care not to drag the instrument needles across anything.

Instrument light bulbs are replaced by simply pulling the bulb socket out of the housing from the rear. Push down and twist the bulbs to remove them from their sockets. The instrument wiring is by simple push-on connectors. The connectors are color-coded, and the instrument housing is stamped with the wire colors for that instrument. All very neat and handy, and a breeze to wire.

As long as the wiring looks original, uncut from previous repairs and not damaged by kinking or overheating, it is probably OK. A simple damp-cloth cleaning will do the most good. Exceptionally dirty wiring can be cleaned with waterless hand

Then, undo wiper switch with an Allen wrench.

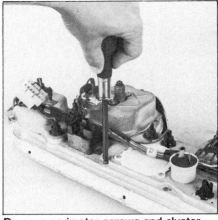

Remove perimeter screws and cluster lifts apart. Replace plug-in turn-signal flasher now if it was faulty.

cleaner.

Many times, however, the wiring will have been repaired by splicing into the loom. This happens after a fire or for installation of auxiliary instruments. If you find a spliced or repaired loom, replace it. While sometimes these repairs are adequate, most of the time they are barely functional. Eventually, they are bound to fail, and the instrument cluster will have to come out to install a new loom. Why not install the new loom now and avoid trouble later?

To install a loom, lay the new loom next to the instrument housing with old the loom

still attached. Then exchange each connection *one at a time!* Use new a new clip to attach the wiring to the instrument housing.

The turn-signal flasher is also attached to the back of the instrument housing. To change it, turn and pull it out. If the flasher has been working, there's no need to change it, but you might want to note its position. It's difficult to find when it fails because you can't hear it clicking. The flasher is the round can with two male spade electrical connectors. It mounts above the fuel gage.

When assembling the '67—68 cluster, remember to install the small rubber stand-

Also while in the neighborhood, replace any burned-out light bulbs.

To rejuvinate needle pointers, spray flourescent orange paint on small brush and apply. Slip cardboard under needle to keep paint from dropping on instrument faces.

New bezels work wonders to rejuvinate old clusters.

offs around the instrument holes in the bezel. These little guys come out either with the lens or stay in the bezel. If you are reusing your lens, it doesn't matter if they are still stuck to the lens or not. But if you are using a new lens, make sure you remove all standoffs from the old lens and transfer them to the bezel.

Don't worry about lens or lens-retainer orientation because they are keyed to the assembly with pins. You'll find it easiest to start with the bezel laying face-down on the bench and building it up from there.

1969—70 Models—Two versions of the late-model instrument cluster are used, but both are serviced the same way. One has only a speedometer, tachometer, water-temperature and fuel-level gages. The other arrangement adds an ammeter and oil-pressure gage.

With the cluster on the bench, the back of the printed circuit, instrument voltage regulator and instrument bulbs are visible. If you need to replace the voltage regulator, just unsnap it from the printed-circuit board clips and remove the single retaining screw. A bad regulator can cause erroneous instrument readings.

If all you are after is the printed circuit, it comes off after removing the voltage regulator, all instrument and indicator bulbs and the circuit-board retaining nuts and screws.

But if you want to disassemble the instrument cluster for cleaning or instrument service, *don't* remove the printed circuit from the cluster! Instead, detach the printed-circuit board from the left and right instrument pods. There are three bulbs and two nuts at each instrument pod which must be removed. At the center of the cluster remove the six retaining screws and lift off the center rear housing and the printed board as a unit.

The speedometer and tachometer or fuel gage will come with the housing. If you want to change an instrument, remove the retaining screws from their dial-face sides and lift out the instrument.

Back at the cluster, remove the retaining screws at each of the two end instruments. The instrument mask and lens are now loose, so pull them free. Finally, lift off the insulator on each instrument stud and pull off the instruments, if you need to change an instrument.

Like the earlier clusters, reassembling '69—70 clusters is pretty simple once you've taken one apart. Remember to assemble the cluster while it faces down. This keeps the instrument pointers from dragging across other parts and breaking.

CLUSTER SERVICE

No matter which instrument cluster you have, it is important to remember how fragile the instruments are. A second's inattention can result in a bent pointer, and for all practical purposes, that means the instrument is ruined. The speedometer and some tachometer needles are fairly robust, but the smaller instruments definitely require all the care you can give them.

What instrument pointers respond to is a light brush coat of fluorescent orange paint. Your best bet for finding this color is to look for a spray can in hobby or hardware stores. While you're there, get a small, natural bristle brush as well.

To paint the pointers, spray a blob of paint on a piece of cardboard. Then dip your brush in the paint puddle. Start at the inboard end of the pointer and work outward. Try for the minimum amount of paint possible. Too heavy a coat will run and sag. Also, on the thin pointers, the paint's weight can affect pointer response time. That may seem hard to believe until you've felt how light the pointers are and think about how heavy the paint is way out on the end of the pointer.

Clean the instrument faces with a soft, dry brush. If they are really dirty, a damp rag may be necessary. Do not use cleaners, especially powerful ones like Fantastik, Formula 409, dish detergent or the like. They lift paint, and will smear the white numeral paint over the black background. Also avoid rubbing your finger over the dial faces and any unnecessary handling.

Because the dial faces are sealed under the lenses, there should not be any grease, oil or heavy grit to clean off. A simple dry brush and damp-rag cleaning will do. Also, avoid blowing the instruments with compressed air. This will damage the pointers and possibly the instrument movement.

If your instrument lens looks like the crystal on a "took a beating" Timex, new ones can be obtained. But if the lens is in good shape, handle it only by the edges until ready for cleaning. Clean the lens under running water by hand. Make sure your hands are clean, so you won't scratch

Check clock operation at battery. Most stop working after a few years, let alone 20 years! Hot setup is having your jeweler install quartz movement in your Mustang clock.

the soft plastic. Also use very light pressure to avoid scratches. Avoid rubbing in circular motions, too. A simple, straight up-and-down stroke makes fewer visible scratches in case you are too rough.

Because the lens is plastic, it will clean beautifully with the cleaners used for aircraft windshields. Meguiar's Mirror Glaze is the most popular cleaner. Look at a local airport for a small bottle.

On '67—68 lenses, you may need to drill a small hole in the clock lens. If you have a clock, the setting knob passes through the hole. If no clock was fitted, a small black plug goes in the hole. Transfer the plug from your old lens.

INSTRUMENT PANEL

For the majority of Mustangs, all the work in restoring the instrument panel is removing the switches and instrument cluster, glove box and dash pad so the metal sections of the dash can be painted. All other instrument-panel parts need a good eyeballing for wear and damage. Many of these parts have already been taken off, like the switches. Others, such as underdash wiring, heaters, A/C systems, defroster ducting, dash vents and wiring looms need nothing more than a good look. These parts are well-protected inside the passenger compartment. Dust, dirt, grease, vibration and sunlight have little effect on them because they are so well-protected. Therefore, in the majority of cases, it is better to leave these parts alone, rather than take them apart just to look at them. Taking

things apart just to take them apart scatters parts—you know what I mean?

Sure, there are a few parts that need attention. Disassembling and cleaning the clock is a logical step now, as is fitting a new lock cylinder in the ignition switch. Most locks are worn out after 20 years, so buy a new matching set for the doors and ignition. The trunk lock can be left alone because it uses a separate key and doesn't get as frequent use as the driver's door and ignition switch.

When examining under-dash wiring, look for worn insulation and hot spots. Check for cuts into the harness where a tape deck or accessory gages were added. If none are found, leave the wiring alone. Under-dash wiring looms can be very expensive. You've got an original Ford part in there now. There's no need to install a replica unless the original has been mauled.

On the other hand, if the wiring is bad, by all means replace it. The last thing you need is an electrical fire in your restored car. You are better off farming out this work while the dash is completely apart, unless you have some electrical background. Connecting a new loom, especially on 69—70 models, takes patience and really isn't much fun. Wire color codes don't necessarily match. So unless you have the time and skills to trace each wire with a test light, leave instrument-panel wiring to someone who does it for a living.

Switches should get a thorough cleaning, or replacement if they aren't working. Radio dials and push buttons respond to plastic cleaner. Meguiar's Mirror Glaze is a good example. Worn out glove boxes can be replaced with new units on '65—66 cars, or better used examples on later cars. Glove-box doors, unlike the instrument bezel, are normally restored, not replaced. Brighten the chrome with very fine steel wool, and repaint the black areas after masking. The running-horse emblem is riveted to the glove-box door, so you'll have to drill out the rivets to remove it. A couple of pop rivets will do the job at reassembly. Woodgrain glove-box doors can be restored using available woodgrain inserts.

Vents should be carefully cleaned with cotton swabs and water. This is tedious, but there is no advantage to removing the vents from the dash. Underdash defroster and heater ducting, wiring, and so on can be wiped clean, vacuumed and blown with compressed air. Underdash A/C units can easily be cleaned thoroughly because you can set them right on the bench in front of you. Use cotton swabs to get in all the tight spots on the unit's front. Be sure to keep the

freon lines closed with tape or plastic caps while the unit is apart. The usual dust and dirt in the shop can really foul the tiny orifices inside the A/C system. This is especially true if there is any bodywork being done nearby. Dust from sanded body filler can be fatal to most mechanical devices.

The dash pad is a replacement item on all Mustangs. Paint the speaker and defroster grilles that fit on top of the dash pad. To save time and for more consistent results, paint all interior trim simultaneously. For example, the instrument panel, the trim at the base of the windshield and metalwork and trim below the rear window can all be painted at the same time.

1965—66 PANEL ASSEMBLY

Assemble the instrument panel after all wiring, painting and cleaning operations are complete.

Heater—Start with the heater if you removed it. After you get the heater assembly into position, install the four nuts that retain it against the firewall. Two wires connect the instrument-panel wiring harness to the heater, just in front of the glove box. There's also a ground wire which goes to the firewall. It attaches with a screw.

From inside the engine compartment, attach the heater hoses. The lower hose is the inlet. Its hose is the one which comes from the intake manifold on V8s or the thermostat housing on six-cylinder cars. The upper hose nipple is the outlet. Its hose goes by the choke on V8 carburetors. On six-cylinder cars, the hose goes to a fitting directly under the carburetor.

Other connections are the fresh-air inlet and defroster hoses, along with the three heater-control cables. The defroster cable attaches to the heater plenum at the extreme left of the heater assembly. The temperature cable goes to the arm atop the heater plenum. This is the one closest to the firewall. The last cable goes to the heater control. It's the fitting near the temperature control, but farther away from the firewall.

Glove Box—After the heater is in, install the glove compartment. Offer the glove box up from under and behind the instrument panel, then screw it in place. This is a tight squeeze, so take your time guiding the fragile, fiber glove box in place. Lay in the glove-box light and attach the glove-box door and retaining strap.

Radio—Next up is the radio. Set the unit in from the back of the instrument panel, making sure the rear support and radio engage. Thread-on the bezel shaft nuts and push on the knobs. Then connect the antenna, power, speaker, pilot light and

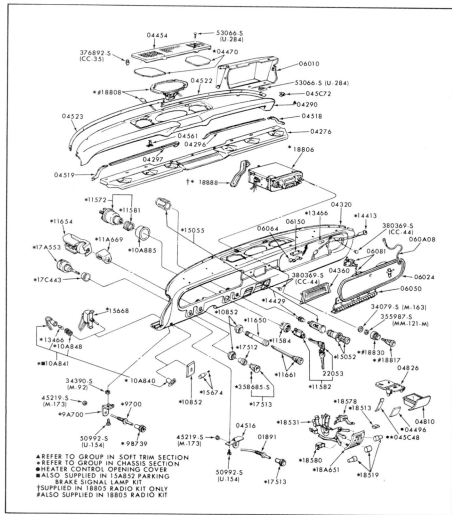

Instrument-panel components: '65—66. Courtesy Ford Motor Company.

Labels in figure:

04454, 53066-S (U-284), 376892-S (CC-35), *04470, 06010, 04522, 53066-S (U-284), *#18808, 045C72, *04290, 04518, 04523, 04276, 04561, 04296, *18806, 04297, 04519, †* 18888, *11572, 04320, *13466, *14413, *11654, *11581, 06150, 380369-S (CC-44), 06064, 060A08, 06081, *15055, *11A669, *17A553, *10A885, 380369-S (CC-44), 04360, 06024, *17C443, *14429, 06050, *15668, 34079-S (M-163), 355987-S (MM-121-M), *10852, *11650, *13466/*10A848, *17512, *11584, *15052, *#18830, *#18817, *■10A841, *11661, 22053, 04826, 34390-S (M-92), *10A840, *11582, 45219-S (M-173), *9700, *15674, *358685-S, *18578, *9A700, *10852, *17513, *18513, 04810, 50992-S (U-154), 45219-S (M-173), 04516, *18531, *04496, **045C48, *9B739, 01891, *18580, 50992-S (U-154), *18A651, *18519, *17513

▲REFER TO GROUP IN SOFT TRIM SECTION
*REFER TO GROUP IN CHASSIS SECTION
●HEATER CONTROL OPENING COVER
■ALSO SUPPLIED IN 15A852 PARKING
   BRAKE SIGNAL LAMP KIT
†SUPPLIED IN 18805 RADIO KIT
#ALSO SUPPLIED IN 18805 RADIO KIT

ground leads.

Switches—Before installing the instrument cluster, fit all switches to the instrument panel. The ignition-switch center section goes in with the key installed. Turn the key to the accessory position and slide into engagement with the outer section. If the lock hangs up, insert a paper clip into the small hole below the key hole. This releases the catch and makes installation easier. The windshield-wiper knob screws onto its shaft from the bottom. The headlight-switch knob merely pushes into position.

Instrument Cluster—Lay a thick cloth across the steering column before installing the instrument cluster. This will prevent scratches to both the column and the bezel.

Lay the cluster on top of the column and connect the wiring. Follow the nearby drawing if you need help. Once you get the wires connected, lay the cluster into the dash and install the screws.

Dash Pad—Finally, top the dash off with the dash pad. If you haven't cleaned the old dash-pad glue off yet, wipe or scrape it off the metal dash section now. Lacquer thinner is a big help here. Once you have all the old glue off, lay a bead of 3M Weatherstrip Adhesive along the same path as the old adhesive. Lay the pad in place and install the two nuts above the radio from behind the instrument panel. Then add the lower trim under the bottom edge of the dash pad. Up top, drop in the defroster ducts, then install the speaker, grille and molding

along the bottom of the windshield. From underneath, hook up the defroster ducts to the heater.

Rally-Pac—If you have a Rally-Pac, install it now. Set the unit on top of the steering column, then attach the lower half of its mounting clamp on the bottom half of the steering column. Tighten the clamping screws while making sure the Rally-Pac is level on the steering column. Use the flags you made earlier as reference when making the electrical connections under the instrument panel.

A/C—If your Mustang is equipped with an underdash A/C unit, hang it by its two mounting bolts. You'll have to pass the freon lines through the firewall while getting the unit into position.

1967—68 PANEL ASSEMBLY

Unfortunately, the first thing that is installed on the '67—68 instrument panels is the dash pad. This cuts off a lot of light under the dash, of course, but is necessary because a multitude of fasteners retaining it are hiding behind the instrument panel.

Dash Pad—Get the new dash pad in position and install the four screws at the forward edge of the pad. At the left edge of the instrument panel, work through the heater-control hole to install the nut on the dash-pad stud. Do the same on the right edge of the dash, working through the glove-compartment hole.

Next, screw on the center and upper finish panels. These are the trim pieces above the radio and glove box. If you have A/C, the center panel is an air outlet. After those two panels, install the right finish panel. It runs down the right side of the glove box. Screws and nuts are used on the backside of the panel.

Heater—From under the dash, remove the nuts on the upper defroster outlets and let the hoses hang. Next route the heater hoses through the firewall, leaving a generous length inside the passenger compartment. Now get the heater assembly into the car and hook up the heater hoses. The top is the outlet; the lower the inlet. The inlet comes from the intake manifold on V8s and from the cylinder head above the thermostat housing on sixes. The outlet returns to the water pump on all engines.

After the hoses are hooked up, bolt the heater assembly to the firewall. It helps if someone can pull the heater hoses from the engine-compartment side of the firewall while you push the heater assembly into position. Now you can get the defroster hoses and heater-control cables connected. See the '65—66 section for control-cable orientation.

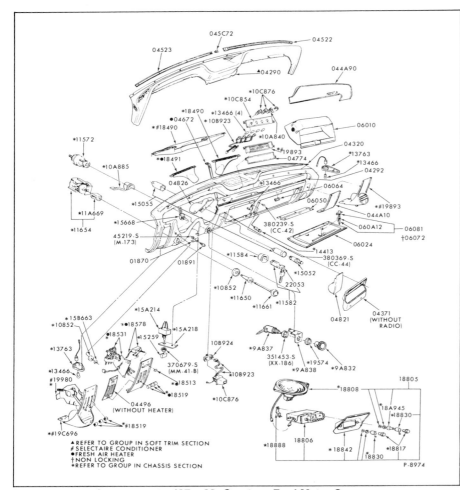

Labels within diagram:
045C72
04523
04522
04290
044A90
10C876
10C854
06010
18490
13466 (4)
10B923
04672
18490
10A840
04320
18490
11572
19893
13763
18491
04774
13466
10A885
04826
13466
04292
06064
15055
06050
06024
11A669
19893
15668
044A10
11654
380239-S
(CC-42)
060A12
06081
45219-S
(M-173)
06072
01870
01891
11584
14413
15052
380369-S
(CC-44)
10852
22053
11650
11661
11582
15A214
04821
04371
(WITHOUT
RADIO)
15B663
10852
18578
15A218
13763
18531
15259
10B924
9A837
13466
351453-S
(XX-186)
19574
19980
370679-S
(MM-41-B)
9A838
9A832
10B923
18808
18805
18513
18519
10C876
18A945
18830
04496
(WITHOUT HEATER)
18888
18806
18842
18817
18519
18830
19C696
▲ REFER TO GROUP IN SOFT TRIM SECTION
SELECTAIRE CONDITIONER
● FRESH AIR HEATER
† NON LOCKING
* REFER TO GROUP IN CHASSIS SECTION
P-8974

Instrument-panel components: '67—68. Courtesy Ford Motor Company.

If you removed the heater controls and cables, install them now. See the drawing for heater control-cable routing. Make sure they go through the hole in the dash brace above the steering column. This supports the heavy cables and keeps them from dangling down against the steering, brake and clutch pedals. Finish off the connections with the electrical leads at the heater assembly and hose connections in the engine compartment.

A/C-Heater—With factory-installed air conditioning, the situation is different because the heating and air-conditioning units are joined in one housing. This adds bulk and connections, and makes installation considerably more involved. As long as you can stand on your head and thread bolts single-handedly using a mirror, you shouldn't have any problems.

Get the defroster and cross-dash A/C ducts in first. You may also want to install the control panel at the left end of the instrument panel and lay the control cables in place. Or if you don't like having the cables dangling under the dash just yet, leave the control panel out until the major components are just about ready to bolt up to the dash.

The big pieces go in roughly from the center of the dash to the right. The heater core, water valve, framework, and temperature-blend door go in first. The A/C evaporator is next, with its brackets and capillary tube. Next on are the 12 case-flange clips, followed by the blower housing and motor. At this point, make as many vacuum and electrical connections as you can, following your numbering system. Follow them with the evaporator case, its

bracketry, the drain tube and more wiring and vacuum hoses. Don't forget the blower-housing mounting stud nut and the two evaporator-case mounting stud nuts in the engine compartment. Finish up with the vacuum supply tank, firewall seal, heater hoses and freon lines. These are only the highlights of A/C installation. The exact order you install this complicated system is variable.

If your Mustang has a convenience control panel, install it before the radio so you'll have maximum working room.

Radio—Radio installation depends on whether your car has a center console or not. If you have a console, install the radio, connecting the rear support and all wiring, but leave the bezel and knobs off. They go on *after* the console is in.

To install the radio, get it under the instrument panel so you can hook up the antenna, power and speaker wires. Then lift the radio up under the dash and slide it into position. Install the mounting screws from in front of the dash and the support at the radio's rear.

Switches—Continuing to the left across the instrument panel, install the ignition and headlight switches, then slide in the ashtray. To install the ignition switch, insert the ignition key, turn the switch to the accessory position and push it in.

Cluster—The instrument cluster is one of the easiest installations on the '67—68 panel. Drape a heavy cloth over the steering column to avoid scratches, then set the cluster up on the column. Join the two multi-pin connectors at the left of the cluster, then lay the cluster into the panel. Install the three screws along the top of the cluster and one in each lower corner. These screws attach to Tinnerman nuts clipped to the panel. So if the screws aren't threading into anything, pull the cluster back out and look for the Tinnerman nuts.

Console—Last on is the center console. Get it approximately in position, and join the electrical connector. Make sure the radio is wired up at this time, as well. Another check is for the anti-chafe strips at the console's under-dash radius. When everything is in place, slide the console into position and install the six attachment screws—three on each side of the drive-shaft tunnel. Finish up with the radio fascia, radio bezels and knobs, shift-lever opening cover and pad. Don't install the center console with the right seat in place. You need the working room the missing right seat gives you.

1969—70 PANEL ASSEMBLY
One advantage to having such a large

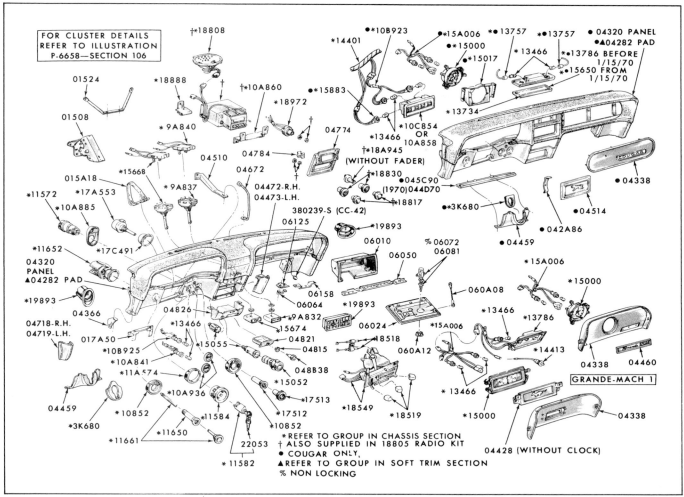

Instrument-panel components: '69—70. Courtesy Ford Motor Company.

Within the diagram:

FOR CLUSTER DETAILS REFER TO ILLUSTRATION P-6658—SECTION 106

* REFER TO GROUP IN CHASSIS SECTION
† ALSO SUPPLIED IN 18805 RADIO KIT
● COUGAR ONLY
▲ REFER TO GROUP IN SOFT TRIM SECTION
% NON LOCKING

GRANDE-MACH 1

04428 (WITHOUT CLOCK)

dash pad on late-model Mustangs is that when it is off, there's lots of room to work on the instrument-panel innards.

Heater—The heater or heater-A/C assembly goes in first. Most cars have only a heater, and it's definitely easier to install. Get the heater assembly under the dash and in position. Working inside the engine compartment, thread on the five nuts that hold the assembly to the firewall. Also install the fan-motor ground wire and heater hoses. The lower hose nipple is the inlet, and its hose comes from the intake manifold. The outlet is the top nipple, and its hose goes to the water pump.

Inside, install the sixth heater-retaining nut on the support bracket. Next install the power-vent air duct—the horizontal section that connects the heater to the defroster duct. This is also the time to connect the

vacuum hoses, following your numbering system. The air-distribution duct can go on now, too. With the hoses joined up, install the heater-control panel in the front of the dash. Depending on how you took the control apart, you'll have several vacuum and cable connections to make at the control or the heater assembly. Now, replace the right instrument-panel brace and its hardware.

When finished with the heater, install the glove box, its door and light.

A/C-Heater Assembly—Just in case you removed the A/C system, here are the highlights of its installation. Start by passing the freon lines through the firewall. Join the blower to the evaporator housing, then get the evaporator onto the dash. Don't tighten the evaporator upper-rear support bracket just yet, although you can wrench down the three nuts on the engine-compartment side

of the firewall. Now you can tighten the upper-rear support bracket. Make the vacuum and electrical connections to the water valve, air doors and so on, following your numbering system.

You can also install the blower-housing-to-firewall screws, and the hose seal and retainer on the engine side of the firewall. Hook up the heater hoses. The inlet coming from the intake manifold is the center nipple. The outlet hose goes to the water pump and is the nipple on the left. Install the freon lines and heat shield over the expansion valve. Hook up the control cable to the temperature blend door.

Because the metal section of the instrument panel had to be undone to service the A/C, align the metal section and install the hardware. The A/C assembly attached to the panel with four bolts, and the steering-

157

column-support hardware can go back on. Snug all hardware, align the instrument panel, then tighten the hardware and check the fit. Don't be afraid to loosen everything and try again if you don't like the fit. After the panel is aligned, install the steering-column lower trim piece.

Back under the dash, install the plenum and make its vacuum and air-duct connections. Finally, install the right instrument-panel brace and feed the evaporator drain tube through the hole in the floor.

Once you've finished all of this, you can rest assured you've just accomplished about the worst mechanical job a Mustang can throw at you. Be sure to take the car to an A/C pro when you are finished for charging the system. He can double-check your work as well as evacuate, leak-test and charge the system with freon.

Radio—If you removed the radio's rear support, install it now. Then hold the radio up to its opening in the dash and connect the antenna, power and speaker leads. Slide the radio into the dash, making sure the rear support goes onto the reinforced section of the dash. You'll have to check the rear of the radio and its support by hand to make sure. At front, install the mounting plate, then front trim panels, bezels and knobs.

Switches—After the radio, replace the ignition, headlight and wiper switches on the instrument panel. If your car was built after January 1, 1970, the ignition switch is in the steering column, so you don't have to worry about it now. Install the outer sections of the switches first, tightening the bezels with a tight-fitting slot-head screwdriver, or a special bezel tool.

Instrument Cluster—Next up is the instrument cluster. Lay a thick cloth over the steering column and set the cluster on it. Connect the wiring multi-connector plug, plus one extra wire to the tach, if so equipped. If there is enough slack, thread on the speedometer cable now. If not, you'll have to wait until the cluster is installed in the dash, and connect it from underneath.

When the cluster is all hooked up, slide it into the dash and install its retaining screws.

Dash Pad—Finish the instrument panel with the dash pad. Before trying to fit the pad to the dash, check that all spring nuts are in place. They go on all the screw holes along the top of the dash pad, at the lower ends, and the center section where it drops down above the radio. When the nuts are all set, lay the pad in position. If your dash has a clock, hook up the clock wires before dropping the dash pad all the way down. If there isn't enough room, drop the pad all the way and hook the wires from under the dash.

With the pad positioned, install its screws. There are three screws along the top, two at the center above the radio, three under the passenger-side hood and three under the instrument-cluster hood. At each end there are two screws in the lower end caps. One faces outside, the other is very low on the end cap.

Console—If equipped with a console, install it now. But first, remove the passenger seat, if not already done. With the right seat out, there is a lot more working room. More elbow room means less chance of breaking the fragile plastic console on the way in. And you don't want to break the console because *factory replacements are no longer available*. Also, make sure the shift knob is off the shifter. You'll need to move the shifter around so it provides maximum clearance to mount the console.

Carefully guide the console into position, moving the shifter as necessary and hooking the electrical leads as you go. With a Shelby console, there are connections to make at the instruments at the dashboard end. Once the console is in, install the retaining screws. Remember to look in the glove-box portion of the console for the screws in there. Finish the console with the trim panel and shift knob.

12

Convertible Top & Vinyl Top

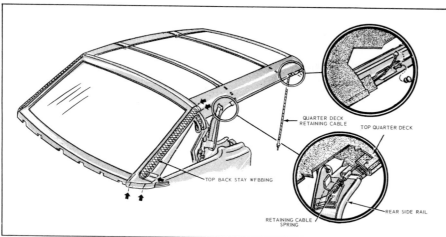

Compared to the ham-fisted, fender-banging chassis and bodywork up to now, convertible and vinyl-top work seems more the domain of the tailor or shoemaker. Here, tacks, staples, glue, cables, cardboard and fabric strips do the job. Courtesy Ford Motor Company.

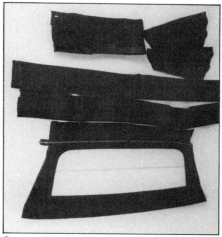

Convertible-top kit with split-glass rear window for '70 project car came from Acme Headliner. Always select a reputable manufacturer who will stand behind their product.

As with the rest of the car, restoring the convertible top seems like an overwhelming job when first viewed. And, as before, you find that it's not so bad when taken in logical steps. The first step, of course, is to remove it from the car.

But first, let's talk about a few definitions before you start taking things apart. What would be the backlite, if the car were a notchback or fastback, is referred to as the *rear curtain*. This may be made of clear vinyl ('65—66) or glass ('67-and-later models). The glass curtain is hinged horizontally so it may fold when the top is down.

Beneath the top, on each side, running front to rear are two *top pads*. These give a smooth line to the curved edges of the top.

Across the rear bow, on the outside of the top, is a separate piece called a *wire-on*. This allows you to cover the staples that would otherwise be visible along this bow. *Bows*, by the way, are the three pieces of the frame that cross the car from one side rail to the other. The first bow is called the

front header.

The *tacking strips* are strips of vinyl or weatherized coils of paper inserted into metal frames upon which the top, pads and rear curtain are stapled or tacked.

With this in mind, let's look at the old top. The first thing to go is the rear *curtain*. Clear vinyl windows are susceptible to the ultraviolet rays of the sun. It becomes brittle and discolored after about two years in the weather—especially in the hot, dry southwestern U.S. This life can be extended by the use of a car cover, preventing those damaging rays of the sun from reaching the rear curtain. Generally, because of the labor involved in replacing the rear curtain, the entire top is usually replaced at the same time.

Next to go is the factory tacking strip found under the front header. This should be replaced with a vinyl tacking strip available from the company from which you purchase the new top. Check this tacking strip carefully and replace it if there is any question.

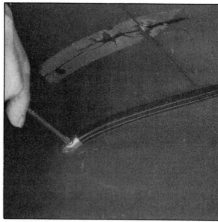

Begin removing top by removing wire-on across rear bow.

CONVERTIBLE-TOP REMOVAL

Convertible-top removal begins with the removal of the weatherstripping around the door and quarter window. Each section of

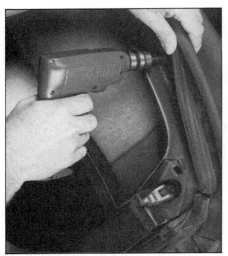

Lay top back into well and remove weatherstrip.

Elevate top a bit to get to side weatherstrips that are part of front seal.

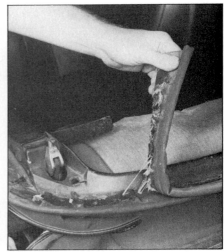

With top folded back, remove front windlace.

weatherstripping is retained to the convertible top frame with nuts and bolts. Remove the nuts with a 3/8-in. wrench or socket. Sometimes it helps if you relieve the tension on the top by raising it about 1—2 ft and placing a box between the front bow and windshield header.

After you remove the weatherstrip, you'll find two retaining tabs sewn to each side of the top. These hold the top in place just aft of the front bow and down the rear side rail behind the quarter window. The tabs are glued in place and should be removed by pulling them away from the side rails. New tabs will be sewn into the top when you receive it.

Now you have enough room to remove the quarter-deck retaining cable. This cable passes through the edge of the quarter deck and is attached at the rear side rail and front side rail, just behind the front bow. Release the retaining-cable spring at the rear, draw it through the top and unhook it from the front side rail. When this is complete, you can release the rear edge of the top around the back curtain and rear-quarter deck.

The first step will be to remove the *top well*—it's behind the rear seat. This is retained with Phillips head screws around the outside of the tacking strip. With it out, you have access to the sheet-metal screws that hold the back-curtain tacking strip to the car body. Remove these and the back curtain and quarter deck will be free.

If you are working with an original top, it will have listings at bows #2 and #3. Notice the screws along the bottom of these

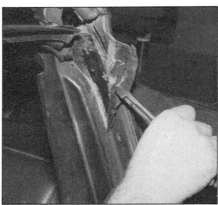

After removing weatherstrip at quarter window, pull up flap glued to top frame.

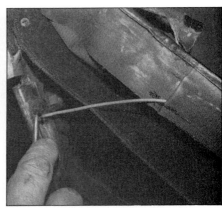

Remove cable that keeps edge of top tight when top is in closed position.

two bows. These screws pass through the listing into a listing retainer. Remove the screws and the listing and retainer will be free. In many aftermarket tops, these listings are not attached to the top. Rather, they are a separate feature that provides a smooth appearance over the top of the bow.

Next, raise the front bow up until the top is perpendicular to the car. With the front bow in this position, you can remove the front weatherstripping, windlace and the front edge of the top from the tacking strip. There remains only the #4 bow outside molding, or *wire-on*, to remove.

Let the top down and move to the molding along the #4 bow. Remove the staples

holding the molding to the top. I like to use an old-fashioned ice pick to start prying out the staple. When you have the staple started, finish pulling it out with a pair of side cutters or wire cutters.

Now remove the staples holding the top of the back curtain and the quarter deck to the bow. When this is done, you can remove the old top from the top frame. Note the tacking strips are still attached to the back curtain. Leave these in place until you can work on it at the bench.

If at all possible, leave the back curtain and sides of the top intact. The old back curtain and sides can be used as patterns to install the new top and curtain.

Move inside car and remove screws retaining tacking strip to body.

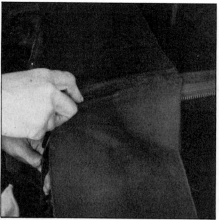

With top off, remove top pads. Measure location of each bow in relation to front bow and body line at rear of top. This will help get bows back into correct position when new pads and top go on.

To remove top bows from body, begin by removing these three bolts (arrows).

Lay the top out on the bench and remove the staples from the tacking strip. Mark each of the three strips right, left and center. Store them away with the old top.

The only things remaining on the frame are the top pads. These are removed by taking out four screws from the front and removing the staples at bows #2, #3 and #4. Before you do this, however, location of the bows must be marked. This can be done in several ways.

Fabricate a bow-alignment gage using the pattern, page 164. This works well for frames that have been disassembled. If you are simply replacing the pads with new ones, remove the old pads one at a time, using the opposite side as a guide for installation. A third solution is to run a length of mechanic's wire from the front bow, wrapped around each consecutive bow (in its correct location) and anchored tightly at the body. This will allow removal of both pads at one time, yet keep the bows in correct alignment.

Top-Bow Reconditioning—You may now want to recondition the top bows if they have lost their luster. Or perhaps, you'd like to rechrome the pins. After several years, many of the plastic bushings may also be worn. Often, as with the demonstration car, the front bow is severely corroded. Any of these will warrant at least a partial, if not complete, disassembly.

Unless you are doing a simple repinning job, the frame must be removed from the car. If you are only repinning and adding new bushings, the top may be left on the car

and intact. Remove the pins one at a time and replace them with a loop of wire. As of this writing, these pins were not available through Ford or through reproduction sources. Therefore, handle them with care. Don't damage or lose them.

Frame Removal—To remove the top frame from the car, release the pressure on the hydraulic rams at the rear by releasing the front hold-down latches. Remove the pin retaining the piston pushrod to the rear frame rail. Remove the three bolts that secure the rear frame rail to its mounting location. Lift the rear frame rail and hydraulic piston as a unit. Remove the two bolts that hold the main-pivot extension plate. This releases the hydraulic piston from the frame rail. Repeat this operation on the other side of the car and remove the frame.

Lay the frame out on a flat surface. Using your Polaroid camera, shoot a whole pack of film of the various connections, unions, frame-member locations, position of pins, and anything that will help you remember how the frame goes back together. If you don't have a Polaroid camera, draw pictures. You needn't be an artist. Simple line drawings will do. In addition, you may wish to label parts for identification and location.

Frame Disassembly—The frame comes apart in a rather forthright manner. You will, however, need one special tool. Because of rust and corrosion, an impact screwdriver is a must. If you don't have one, its minimal cost makes it an excellent

Remove pin from ram to free top bows from body. They may then be removed as a unit.

investment. As always, buy the best.

Begin disassembly by removing the hold-down latches at each front corner. Remove three small bolts and the large bolt that holds the clamp handle to the frame. Next, remove the two bolts on each side of the front bow retaining it to the front side rails. A little encouragement with a hammer may be necessary. Remember, you're working with castings that break easily. Therefore, don't get carried away with your "encouragement."

With top bows disassembled, there's a lot of parts to keep track of. Be sure to mark each piece, make drawings and take photos as you go. I use an impact screwdriver to loosen rusted latch-lever bolt. Separate parts, if necessary, by tapping with large, rubber mallet.

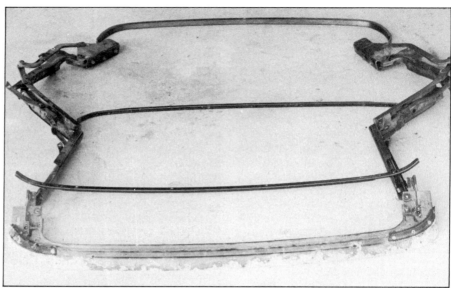

After removing bows, side rails may be disassembled.

The next step is to remove the three remaining bows. These are pinned to the side frame rails. Remove the hairpin clips and drive the pins out with a drift punch.

This leaves only the side frame rails to contend with. These are both screwed and pinned in position. Some of the Phillips head bolts may require the use of your impact driver to loosen. Separate each frame rail from its partner.

Remove the remaining levers on the frame rails and the bow attaching brackets from the bows. Prepare all pieces for painting or powder-coating. If you elect to powder-coat the pieces, be sure you indicate on each piece where it's *not* to be coated. This includes all pivot points. If these bores are coated, you will have a difficult time trying to get the pins and bushings back in!

Front-Bow Reconditioning—The front bow supports a great deal of the top. With it is a tacking strip, weatherstrip and windlace. Generally, the front bow itself is in terrible condition, having weathered several years.

The tacking strip is on the underside of the header holds the front of the top in place. A rubber weatherstrip, retained by screws is mounted directly behind the tacking strip prevents wind and rain from flowing in between the header and windshield

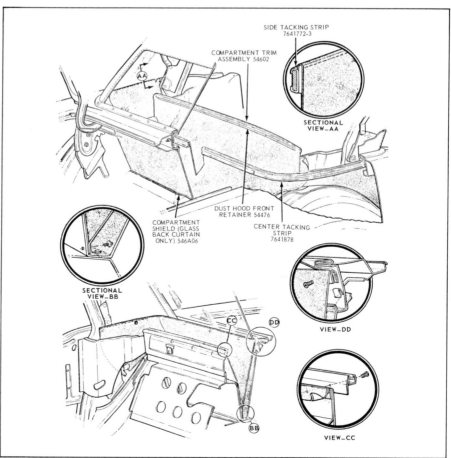

Typical back-curtain installation. Courtesy Ford Motor Company.

top molding. The windlace is a 1/2-in. round rubber bead covered with top material and stapled to the tacking strip. This cosmetically seals the gap between the header and top windshield molding.

The windlace is not included in the top kit. Remove the rubber core form the old windlace and take it to your local trim shop along with the 4-in. X 6-ft. strip of top material *included* in the top kit. Here, they will cover the 1/2-in. rubber core with the new material using their big sewing machine. The cost should be less than a couple of dollars.

The windlace and weatherstrip were removed when the top was removed, leaving the tacking strip. This is removed by drilling out the rivets securing it to the bow. If, as on the demonstration model, the screws holding the front of the pads are corroded to the bow, they too, must be drilled out. When everything has been removed from the bow, send it out to be sandblasted or beadblasted.

If extensive pitting has occurred, these pits may be cosmetically repaired with plastic body filler. Fill, file, sand and paint. This will give you a smooth surface under the new top.

A new, 3/16 X 1/2-in. vinyl tacking strip may be purchased from the same dealer who sold you the convertible top. To install it, use #6 X 1/2-in. pan-head Phillips screws. These will come through the top of the bow. Use a body grinder to grind them flush with the surface of the bow. Spot-paint the area to retard corrosion.

Frame Assembly—After painting or coating all parts, you can begin reassembly. This is basically the opposite of disassembly, but there are a few things to keep in mind. All pivot points had plastic bushings except at the pin between the front side rail and the intermediate side rail.

Replace the lever assemblies to the front and intermediate side rails before connecting the side rails.

Assemble the side rails first, then bows #4 through #1, finishing with the front latch assemblies.

When the frame is completely assembled, return it to the car. Bolt it in place after mounting the hydraulic pistons. Now the top must be adjusted to fit centrally within the car.

Loosen the three screws you just secured to mount the main pivot bracket to its support. On 1965—68 models, shift the main pivot bracket toward either side as necessary to obtain a clearance of 9/32 in. between the rear side rail and the quarter outside-rear side-belt moldings and then

After painting and assembling bows, mount them back into car according to measurements you took earlier. I use seat-belt webbing, clamps and staples to hold bows in location while I install pads.

secure the bolts. On 1969—70 cars, this clearance is 7/16 in.

There are two side-rail adjustments plus toggle-clamp and dowel-pin adjustments to be made. These are best left, however, until the top pads are installed.

CONVERTIBLE-TOP INSTALLATION

Top Pads—Before starting the installation of the top, decide on the tools you'll use to attach it to the frame. The tool of preference is the pneumatic staple gun. This may be borrowed from an upholstery shop, if you are on excellent terms with the owner. It's not often found in tool-rental shops. And unless you plan to use it extensively, it's a rather expensive investment for the home mechanic.

Some tool-rental shops have electric staple guns. These will do a very good job if they will shoot staples at least 3/8-in. long. A better size is 1/2 in. You'll need staples of 1/4, 3/8 and 1/2 in. to do the job. An electric staple gun costs about half that of a pneumatic gun and does not require an air compressor to operate.

Before all these fancy tools were invented, coach and auto trimmers managed very well with tacks and a hammer. These are still available today.

If you've never seen a professional auto trimmer at work, here's how it goes. He holds a tack hammer with a magnet on one end in one hand, and places some sterilized tacks in his mouth between his jaw and

gum, just like a chew of tobacco. He then pulls the tacks out, one at a time with his tongue. Finally, he places the tack between his teeth, picking it up with the magnet on his hammer and drives the tack home while holding the material taught with his free hand.

I suggest you don't try this. It's a bit hard to swallow a tack, but very easy to inhale one. The alternative method is to sprinkle a few tacks out on a non-magnetic surface and pick them up, one at a time, with the magnet. This could save a large doctor bill. If you elect to use the tack method, purchase some #3, #6 and #12s.

The installation of a new convertible top is preceded by the installation of new pads. These pads are available as part of the convertible-top kit, or may be purchased separately.

If you have completely reconditioned the top frame, make the bow-alignment jig. Retain it in place with the bows in their corresponding notches. With the top latched down, mount the front of the pad (the narrow end) to the pad location on the front bow and secure it with four countersunk, flathead, Phillips head screws and bezel washers.

Pull the pad tightly across the four bows and staple the back of the pad to the #4 bow. Note the inboard notch in the metal part of bow #4. Align the inboard edge of the pad with this notch. The pad will also extend beyond the bow. After installing both pads and checking for true alignment of bow #4, trim the selvage end of the pads.

To check for true alignment of the rear bow, measure from the bow, at the pad location, to the rear belt molding on the body. Both sides should be within 1/4 in. of one another. Tighten the loose side. Leave the bow fixture in place until the back curtain has been installed.

Bow Adjustments—Two adjustments can be made to the top bows. Both are used to adjust the gap between the door and quarter window glass, and the front and intermediate (center) side rails. One is the eccentric pin and locknut at the rear side rail attached to the balance link and the other is at the #2 bow attaching bracket.

To make this adjustment, raise the door and quarter-window glass to the full *closed* position. Loosen the locknut on the eccentric pin at the rear side rail. Rotate the eccentric pin while holding the locknut. Work for an even fit between the weatherstrip and window glass. Tighten the locknut.

Move to #2 bow and repeat the same

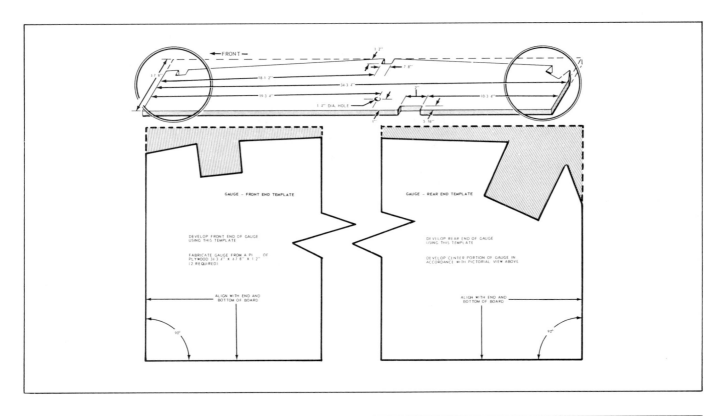

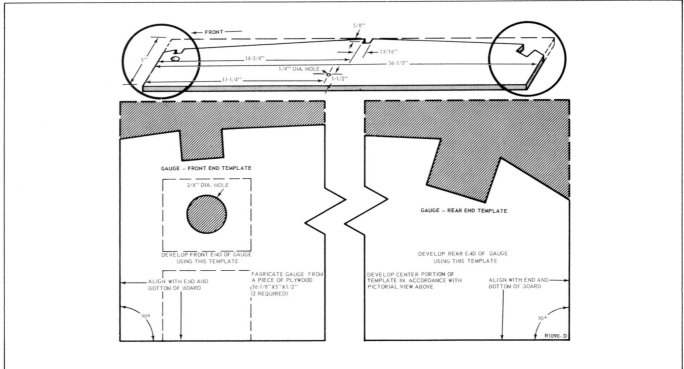

Use these patterns to make bow-alignment jigs for your Mustang: '65—68 models (top), '69—70 models (bottom). Make them from 1/2-in. plywood and hold to frame rails with mechanic's wire. Jigs hold bows in relative position while you make fine adjustments from previous measurements.

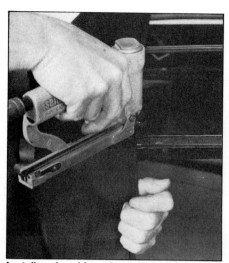
Install pads *without* foam inner piece.

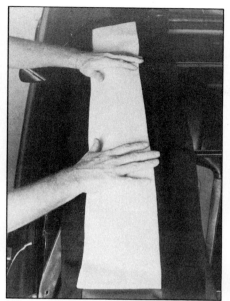
When pads have been stapled to bows, lay foam on pad and fold side pieces over it.

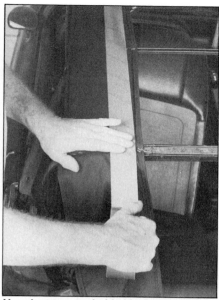
Use duct tape to hold side flaps together over foam. It will not be seen when top is installed.

operation at the eccentric on the bow attaching bracket. Work again to even the space between the weatherstrip and glass.

The toggle clamp can be adjusted by loosening the Allen setscrew and rotating the "J" hook clockwise to tighten and counterclockwise to loosen. This will clamp the front bow to the windshield header tighter or loosen it a bit.

Dowel-pin adjustment is equally simple. Loosen the retaining nut, adjust the pin and then lock the nut.

These four adjustments should give the final fit needed to seal the top against any weather conditions short of a tornado.

Rear Tacking Strip—Open the package your top came in and remove everything. You should have, besides the two pads you've already installed, a back curtain, two rear pads (or belts), a 6-ft piece of wire-on with two decorative end clips, a front windlace and the top itself. The top is made in three pieces: the deck and two side quarters.

Set everything aside except the back curtain. Lay this out on a clean bench, and find and mark a vertical center line at the top only. Then set the old curtain on top. Carefully, align the old and new zippers. Now, mark the new curtain bottom edge for the tacking strip to correspond with the old curtain. Cut out for the bolt holes but do not trim the selvage. Do this with the old and

new tops and the back belt. The idea is to place the tacking strips in the same location on the new top as they were on the old. Don't forget to mark a center line at the front edge of the top deck.

When you're sure of the locations, staple or tack the tacking strips to the back window and bottom side quarters of the top. Be sure the belts are located between the back curtain and side quarters. Now, carry the entire mess over to the car, and place it over the frame. The tacking strip goes down into the top well with the top and quarter decks spread out over the bows. In order to mount the tacking strips, you'll have to get the back curtain out of your way. Temporarily tack it in place, centered on the back bow.

Now, screw the tacking strips to the body in their original locations. This will take a little wrestling because the new fabric will not be molded to shape. The use of a long, 1/4-in. punch will help align the tacking strip with its corresponding holes. Set the bolts as loosely as possible until all have been placed, then tighten them.

Locating Top—After the tacking strip has been fastened to the car, fold the top fabric back, exposing the back curtain and rear bow. Climb into the car, and stand between bows #3 and #4. You're now going to fasten the back curtain and belts to the rear bow.

Remove the tacks or staples you used to

Mark location of tacking strip on rear window, or curtain, and on each side quarter. Then, align old rear curtain with new and transfer locating marks for tacking strip onto new curtain. Repeat process with side quarters.

hold the back curtain temporarily in place. Mark the center line of the rear bow. Match the center line of the back curtain with the rear-bow center line, pull the curtain as tight as you can and place two or three tacks or staples to hold it there. Working from center left, then center right, pull the cur-

Now, staple rear curtain to the tacking strip after first using your locating marks for positioning.

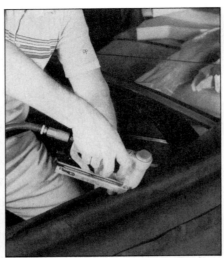

Here, I have bolted tacking strips back to body and am pulling rear curtain tight over back bow as I staple it in place. Repeat process with rear belts.

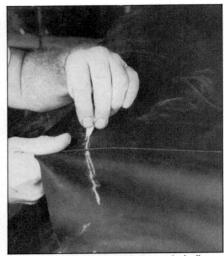

Pull top tight over front bow and chalk a line for its position. Elevate top and staple fabric to front bow using chalk line.

the curtain as tight as possible and tack or staple in place. Do not trim the selvage.

Pull the belts into place, being sure there are no wrinkles, and staple them to the back bow. The back curtain is now in place and should be tight with no wrinkles. Again, lay the top out over the bows. You may now remove the bow locating gage.

Go to the front of the car and align the center line of the top with the center line of the front bow. Pull the top gently and clamp it in place with a spring clamp. Go to either the right- or left-quarter-deck seam and pull it tightly, clamping it in place also. Repeat this on the other side.

Now go back to the rear bow. There is a small (2 or 3-in.) seam located at each end of the long opening between the top deck and the zipper guard. This seam, called the *outboard bond seam*, should be directly over the fourth bow and there should be few, if any, wrinkles in the quarters. If the seam falls in the right place, over the bow, you're in luck. Jump ahead and finish the job. If the seam is off, it must be corrected.

If the seam is off by no more than 3/8 in. in either direction, it can be remedied from the outside. If it is off much more than this, you've done something wrong. Go back and check your work. Remove the top, make the necessary corrections and install the top again.

If the seam is off less than 3/8 in. and to the *rear* of the bow, try pulling harder on

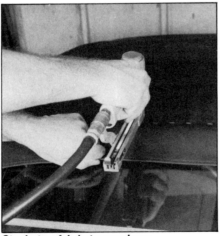

Staple top fabric to rear bow.

the main seam between the deck and quarter. You can pull like crazy here without worrying. This will probably pull it into place. If it doesn't, there is yet another trick. Unlatch the top, force the rear bow back until it's under the seam and then tack or staple the area at the seam to the bow. Then latch the top down again. This will only work if the seam is off no more than 3/8 in. If you try to do this with an outboard bond seam that's 1-in. off, you'll tear the fabric.

If the outboard bond seam is *forward* of the bow, tack or staple the area behind the seam in place. Release the clamp at the front seam, pull the bond seam back in place over the bow and staple the material in front of the bond seam to the bow. Now you have the bond seam puckered up above the bow. It may then be cut away with scissors or a razor blade. Later, it will be covered with the wire-on. Finish by pulling the quarter-deck seam tight again and clamping it in place.

When the bond seam falls into place, staple it and the quarter-deck seam to the rear bow. If the top deck is smooth and taught, go ahead and staple or tack it down across the rear bow. If it is not taught, stretch tighter on the quarter-deck seams, making it taught.

Now, position the flap that covers the zipper and staple or tack it in place. You now have the back curtain, zipper flap and top deck all fastened to the fourth bow. With this, the rear of the top should be smooth and wrinkle-free. If there are any wrinkles, you should be able to remove them by pulling gently on the retaining tabs sewn into the top near the rear frame rail. Move to the front bow, get out a pencil or chalk and follow along. You're going to pull the top deck tight from the center both ways. As you pull it tight, mark the outline of the front edge of the front bow on the outside of the top fabric. Do this all the way

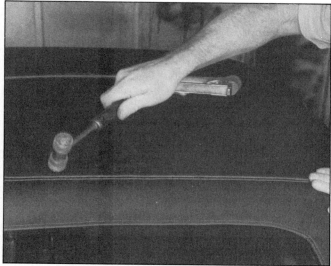

To cover these staples, use wire-on included in kit. Lay wire-on over staples, pulling it tight. Staple it down, fold edge over and flatten it with a plastic or leather mallet.

across the front of the top.

Unlatch the top and raise the front bow until it is vertical. While standing on the front floor pan, pull the top material around the front bow until the center line and bow edge line you just drew align together at the front of the bow. Place a tack or staple. Carefully continue tacking or stapling the front edge of the top fabric to the front bow, making sure the edge line you drew follows the edge of the bow. Finish off by folding the outside bound edge under and staple or tack it in place.

Lower the top. Latch it tight and see how the front edge fits. If there are any loose spots, mark them, raise the front bow and adjust the stapling or tacking to remove them. If the top fits to your satisfaction, install the quarter-deck retaining cables.

Cut a piece of mechanic's wire about 5 or 6-ft. long. Make a small loop in one end and bend about 1 in. of the other end back over itself. You have now formed a very long needle with a smooth, blunt end. Thread this through the quarter-deck channel that houses the retaining cable, working from the front of the top to the rear. Hook the retaining cable spring into the loop you made and thread it through the channel. Connect both ends of the cable and set the front bow down.

The side tabs should now be glued to the frame rails. Use contact cement to secure them in place. Pull the rear tabs around the

rear frame rail until the edge line of the top is in line with the rear frame rail.

Return to the rear of the car and install the wire-on across the rear bow. It should extend past the bond seam about 4 in. on each side. After tacking or stapling the wire-on in place, fold it over by hand. Then use a small mallet to seal it. Finish the ends off with the two chrome tips and screws included in your kit.

Convertible-top installation will be complete when the weatherstripping, front windlace and top-well cover are installed.

The front windlace is tacked or stapled to the same tacking strip as the front of the top. Follow the photographs to see how the ends of the windlace are treated. If your 1969 or '70 car still has the solid rubber windseal and it's in good condition, clean it, put some rubber preservative on it and install it in place of the windlace. As of this writing, this windseal is no longer available or being reproduced.

This completes the installation of the convertible top. If there are any *small* wrinkles, these can be removed with the heat gun as they were with seat covers. Vinyl-top material, however, does not shrink as much as vinyl seat-cover material. So the wrinkles must be minor. Another help will be to apply clear, cold water to the fabric on the inside, just in the area of the wrinkle. This will also encourage a little shrinking.

Kit includes a 4-in. wide piece of top material about 5-ft long. Remove rubber core from old windlace, take core and new material to local upholstery shop and have them cover core with your material. This is now a windlace and can be stapled or cemented to front bow as shown. Replace weatherstrips and your new top is ready to go.

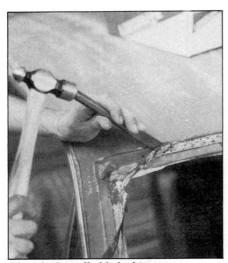

After ripping off old vinyl top, use a solvent such as lacquer thinner to wash away old cement. Be careful not to splash thinner on paint. Remove drive nails at window openings with hammer and chisel.

After removing weatherstrips from door openings, stainless trim on drip rail must be removed. Drill out rivets using a #30 or 1/8-in drill bit.

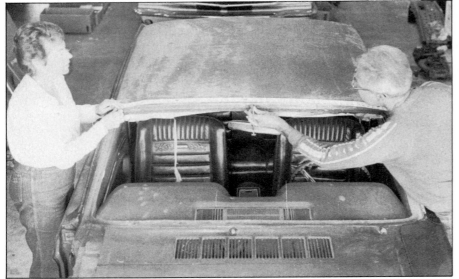

To correctly center top on car, find roof center line and mark.

VINYL-TOP INSTALLATION

As you move along through this book, you may have noticed that there are often two ways to do things, especially where fabrics are concerned. So it is with installing a vinyl top. The correct way to install the vinyl top includes removing the windshield and backlite. But "quickie" shops often leave the glass in, and the job lasts but a few months. If you've waited this long to restore your Mustang, don't take the low road to shortcuts now. See page 41 for a demonstration of glass removal.

Top Removal—Because you may not be doing a complete restoration, I'll describe top removal as if that were the only thing you were doing to the car.

After removing the glass, remove the rear seat and quarter trim panels. This will give you access to the headliner. Loosen the headliner at the rear quarters to reach the retaining nuts on the side belt moldings. Remove the nuts and then the moldings.

The weatherseal around the door must be removed to reach the drip-rail molding. The molding is fixed to the drip rail with 1/8-in. rivets. These rivets pass through the drip rail and are secured by an aluminum strip in the drip rail called the *roof side-cover retainer*. When you drill out the rivets, the drip-rail molding will come off and the roof side-cover retainer will come out. You have one more piece of hardware to worry about before the vinyl can be removed.

At the end of each seam, in the opening of the windshield and backlite, there is a large staple or drive nail. Remove it with a sharp chisel and hammer. Next, lever off all of the windshield- and backlite-molding retaining clips. Be prepared to replace any weak or broken ones.

Now you can tear the old top off the roof. In most cases, the surface of the roof will be left uneven because of the old cement. Clean it off with an inexpensive wash thinner or lacquer. Don't forget to clean inside the windshield and backlite openings, also. Any bumps or lumps left behind will be seen under the new top.

Check for Rust—And if you spot *any* rust, remove it before going any further. Light surface corrosion can be removed with a D-A sander and 220-grit paper. Be sure to follow up with metal prep to *phosphatize* the bare steel before recovering the top.

Vinyl-Top Installation—Remove the cover from the box and throw it over the roof. Center it fore and aft to be sure it is large enough. At the front edge, just along the break between the roof line and the windshield opening, draw a 6-in. line on the *underside* of the vinyl. This will help to position the top when you're ready to cement it in place.

Lift the top from the roof and lay it out on a clean surface. By folding it over on itself exactly in half, you can find its center line. Mark the crease on the *inside* of the top (on the cloth backing) at the front and rear edges. Unfold the top and snap a chalk line connecting the two marks. This is the center line of the top.

By measuring inside the opening of the windshield and backlite, you can find the center of the roof. Again, snap a chalk line on the roof of the car.

To apply an even coat of vinyl-top cement to the roof, it should be sprayed on.

Find and mark center line on *underside* of new top.

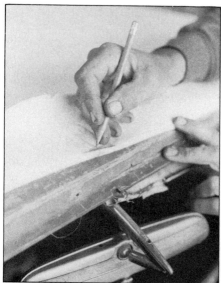

Lay top on roof and match center lines. Then adjust the top from front to rear. Mark top to locate where it will lie, front to rear.

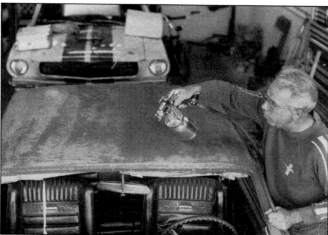

Remove top and run a 1-ft wide band of contact cement along its center line. Do the same for the new top.

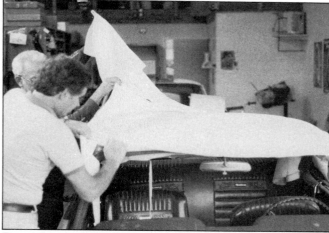

To set the top, enlist the help of a friend. Find center line and marks for front edge. Line these up, stretch top tight and bring cemented sides into contact with one another.

There are two options: borrow a painter's old primer gun, wash it out thoroughly and load it with cement. (Of course, you will clean it equally as thorough before you return it.) The second option is to purchase a couple of aerosol cans of contact cement. The first option is the best.

If option number one is available to you, purchase a quart of *uncolored* contact cement and thinner. If there is a coloring agent in the cement or thinner, it will bleed through the vinyl top. If thinner is unavailable, use wash thinner you bought to clean the top. Check to be sure the lacquer thinner will thin the contact cement before mixing a full quart.

Add enough thinner to the cement to allow it to spray out without clumping or turning to cobwebs. Too little thinner makes the cement lumpy; too much and you spray cobwebs instead of cement. Start with a mixture of two-thirds cement and one-third thinner.

Now things get a bit tricky. Call in a friend to help. Spray a line of cement about 12-in. wide down the center line of the roof. Spray enough to give at least 75% coverage. Do the same to the underside of the vinyl along the center line. Allow the cement to dry until it is no longer tacky. To test for dryness, place a piece of paper on the cement and gently rest your hand on it. If you can pull the paper away without

Fold top back on itself, spray roof and top with a thin layer of cement. Fold top back onto roof, keeping it taught and wrinkle-free as you go.

Relieve stress in corners by radiusing fabric with your shears. Don't cut too far.

Trim excess material *inside* drip rail. Be careful not to cut too close to roof. After you finish cementing roof, apply silicone sealer inside drip rail to keep water from seeping under vinyl and causing rust.

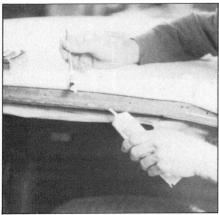

Install reveal-molding clips back in their original holes, install windshield and backlite, then reveal moldings and job is done.

Cement top down to windshield and backlite openings. Use contact cement from a tube or can with a brush. If you try to use your spray gun you'll have cement all over everything.

sticking, the cement is dry.

With your helper, lift the top and carry it to the car. Turn it over so the two cemented surfaces will contact. Remember, *you only get one try! The first time must be right!* If you have to lift the top after cementing it, the cement will ball-up and cause unsightly bumps under the finished top.

Set the front edge first, taking note of the front edge mark and center line. Your partner will match these marks while you hold the top up from the rear. When the front edge is set snugly, pull along the center,

matching the two center lines (top and roof). While maintaining tension, your partner should push the fabric down, cementing the top to the roof. It's going nowhere now. The rest is simple.

Fold one side of the top over and spray cement on it. Cover that side of the roof with cement. Give the cement time to dry. Then pull the side over. Again an extra set of hands helps. Be sure the seam is straight. Repeat this operation on the other side. Finish spraying by cementing the windshield-pillar covering in place. Now

the top is on and you can finish off around the windshield and backlite.

Find that remaining tube of 3M Weatherstrip adhesive, or equivalent, and a couple of acid swabs. Cement around the inside opening of the windshield and a corresponding area of the top. As you bring the two cemented areas together, find the screw holes for the windshield-molding clips. Use an awl to punch these as you go. You'll have a hard time finding them if you don't.

It will help to clip the corners. Be sure you don't cut in too far or it will be visible after the windshield is in. Where each side seam ends, put in a drive nail. This prevents any shrinking of the seam. Finish the windshield area and repeat the process at the backlite.

Locate the holes for the belt molding. Trim the excess top material just below these holes and install the belt molding.

You can now finish off and replace the glass and hardware you removed. When the drip-rail molding and roof side-cover retainers have been installed, run a bead of silicone sealer into the drip rail. Use white sealer with a light top and black with a dark top. This seals the top from moisture at the edges and prevents water from seeping into the car through the rivet holes.

Slide the windshield- and backlite-molding retaining clips into place. Install the windshield and backlite, page 137, and the job is done.

PRODUCTION FIGURES

1964-1/2	Convertibles	28,833
	Coupes	92,705
		121,538
1965	Convertibles	73,112
	Coupes	409,260
	2+2	77,079
		559,451
1966	Convertibles	72,119
	Coupes	499,751
	2+2	35,698
		607,568
1967	Convertibles	44,808
	Coupes	356,271
	2+2	71,042
		472,121
1968	Convertibles	3,339
	Coupes	249,447
	2+2	42,581
		295,357
1969	Convertibles	14,746
	Coupes	128,458
	Grandes	22,182
	Mach I	72,458
	Sports Roof	61,980
		299,824
1970	Convertibles	7,673
	Coupes	82,569
	Grandes	13,581
	Mach I	40,970
	Sports Roof	45,934
		190,727

Shelby

1965	GT350	562
1966	GT350	1442
	GT350H	936
1967	GT350	1175
	GT500	2050
1968	GT350 (2+2)	1253
	GT350 (Conv)	404
	GT500 (2+2)	1140
	GT500 (Conv)	402
	GT500KR (2+2)	933
	GT500 KK (Conv)	318
1969	GT350 (2+2)	1085
	GT350 (Conv)	194
	GT500 (2+2)	1536
	GT500 (Conv)	335

ENGINE COLOR DETAILING

1965	170	Black block and heads, orange valve covers, orange air cleaner.
	200	Black block and heads, orange valve covers, orange air cleaner.@
	260	Black block, blue heads, blue valve covers, blue air cleaner.
	289	Black block and heads, gold valve covers, gold air cleaner.*
1966—68		All engine parts painted medium blue.
1969—70		All engine parts painted light blue.

Exceptions:

a. High Performance 289 had chrome valve covers, air cleaner and oil cap.

b. Engines with dress-up kits also had chrome radiator cap, master-cylinder cap and top of oil dip stick.

c. Shelbys had a chromed air cleaner, aluminum valve covers and oil pans.

@ By late 1965, Ford painted 200s medium blue.

* Intake of air cleaner painted flat black.

color codes: Orr-Lac engine enamels

Black	903	Dark blue	963
Gold	908	Orange	912
Light blue	958		

Following items should be left as natural metal. Beadblast and phosphate or coat with a clear acrylic to imitate factory phosphate finish:

Accelerator and linkage
Air-conditioning line
Air-conditioning-compressor-mounting bracket
Air-cleaner wing nut
Alternator and alternator bracket
Ball joints
Bell housing
Brake-junction block
Brake lines
Carburetor spacer
Clutch linkage
Differential yoke
Differential nuts
Disc-brake assembly
Door locks
Door striker
Drive shaft
Emergency-brake linkage
Engine-accessory-bracket bolts
Engine-support bolts
Exhaust hangers
Exhaust system
Fan spacer
Fender bolts (interior only)
Fuel tank

Fuel-tank-filler neck
Fuel-tank bolts (all)
Fuel pump
Gas lines
Generator brackets
Grille bolts
Hood catch
Hood latch
Hood bolts
Hood hinges
Hood-hinge bolts
Idler arm
Master cylinder
Oil-pressure-sending unit
Parking-light bodies
Power-steering brackets, 1967 and '68
Power-steering-pump brackets
Proportioning valve
Radiator bolts
Leaf-spring shackles
Leaf-spring mounting bolts
Shock-mounting bolts
Shock-tower-brace bolts
Shifter linkage
Steering-gear box
Suspension attaching bolts
Taillight bodies
Taillight-housing nuts
Transmission
Transmission crossmember
Trunk lock
trunk-lock striker
Universal joints
Valance-panel bolts
Voltage-regulator bolts
Wheel-cylinder bolts
Wheel-cylinder bleeder screws
Windshield-washer pump
Windshield-washer-bag retaining bolts

Following items should be painted chassis black:

Battery tray
Battery holddown
Brake backing plates
Bumper braces
Coil springs
Crankshaft pulley
Fan
Fan pulley
Fan shroud
Fender wells
Fenders, inner
Firewall
Front crossmember
Generator
Lower control arms, front suspension
Heater motor
Hood-lock support and brace
Radiator
Radiator support
Rear-axle housing
Shock mounts
Spring shield
Starter
Support braces
Undercarriage
Upper control arms, front suspension

Suppliers List

Acme Headliner
550 W. 16th St.
Long Beach, CA 90813
(714) 437-0061
Headliners, convertible tops, carpets

Tom Adams
14216 N. 39th Dr.
Phoenix, AZ 85023
(602) 978-2955
Accessories, parts, literature

Bill Alprin
184 Rivervale Rd.
River Vale, NJ 07675
(201) 666-3975
NOS parts

Anderson's Mustang Corral
11551 K-32 Highway
Bonner Springs, KS 66012
(913) 441-8103
Used parts

Antique Automotive of San Diego
4124 Poplar St.
San Diego, CA 92105
(619) 283-6626
Parts, car dealer

Arizona Mustang Supply
P.O. Box 12485
Tucson, AZ 85732
(602) 747-9709
New & used parts, restoration

Auto Krafters
6000 Q St.
Omaha, NB 68127
(800) 228-7346, (402) 734-7557
Parts, literature

Auto Salvage Brokers
Route 5, Box 568
Orlando, FL 32807
(305) 275-8721
Comprehensive parts, complete cars

B.S.I.A. Mustang Supply
303 Brighton St.
LaPorte, IN 46350
(219) 326-1300
Parts

Bill's Speed Shop
13951 Millersburg SW
Navarre, OH 44662
(216) 832-9403
Body parts

Boss Hoss Corral
1114 S. 41st
Temple, TX 76501
(817) 773-4000
Used parts

Tony D. Branda Shelby & Mustang Parts
1434 E. Pleasant Valley Blvd.
Altoona, PA 16602
(800) 458-3477, (814) 942-1869
Parts, accessories

C.A.R. Distributors
12375 New Holland St.
Holland, MI 49423
(616) 399-6783
Parts

California Auto Trim
10949 Tuxford Street #1
Sun Valley, CA 91352
Interior work, parts, service

California Mustang Sales & Parts
18435 Valley Blvd.
La Puente, CA 91744
(800) 854-1737, (818) 964-0911
Parts, upholstery, trim

Canadian Mustang
450 Swift St.
Victoria, BC Canada VSW IS3
(604) 385-7161
Accessories, carpets, parts, tops

Capital Auto Repair
1903 Unit E. West Vista Way
Vista, CA 92018
Custom engine rebuilding

The Carriage House
324 W. Nakoma
San Antonio, TX 78216
(512) 344-7358
Restoration, engine rebuilding

Circle City Mustang
Rt. 1, Box 27
Midland City, AL 36350
(205) 983-5450
Parts

Classic & Collectible Cars
4710 S. Arville
Las Vegas, NV 89103
(702) 873-2222
Car dealer, accessories, detailing

Classic Mustang, Inc.
117 Broad St.
Forestville, CT 06010
(800) 243-2742, (203) 582-3191
Parts & accessories

Coach Craft
6798 Mid Cities Ave.
Beltsville, MD 20705
(301) 937-5834, (800) 638-0257
Restoration, floor pans, lacquer
painting, frame members

Coachcraft Ltd.
158 W. Valley Avenue
Birmingham, AL 35209
(205) 942-4202
Bodywork, painting, restoration

Cobra Restorer's Ltd.
3099 Carter Circle
Kennesaw, GA 30144
(404) 427-0020
Parts, restorations

Columbia Auto Restorations
3200 W. Metaline Place
Kennewick, WA 99336
New & used parts, accessories,
appraisals

Colorado Mustang Specialists, Inc.
19000 E. Colfax
Aurora, CO 80011
(303) 343-7036
New & used parts, restoration,
storage

Consolidated Buyers Ltd.
P.O. Box 5391
Garden Grove, CA 92645
(714) 897-0652, (213) 596-0951
Parts, restorations

Corbett's House Of Ford Parts
190 Calhoun St.
Edgewater, MD 21037
(301) 269-1167
NOS parts, shop manuals, parts
books

Council Street Automotive
55 E. Council
Tucson, AZ 85701
(602) 624-3581
Parts, restoration, service

Creative Workshop Motorcar Restorations
3052 SW 4th. Ave
Ft. Lauderdale, FL 33315
(305) 522-1682
Parts, restoration, service

Crossroads Classic Mustang
12421 Riverside Ave.
Mira Loma, CA 91752
(800) 443-3987, (714) 986-6789
Parts & accessories

Dad's Mustang Supply of Maryland
3212 Ascot Lane
Fallston, MD 21047
(301) 692-5515
Parts

Dallas Mustang Parts
9515 Skillman St.
Dallas, TX 75243
(800) 527-1223, (214) 349-0991
Reproduction parts

The Eastwood Company
580 Lancaster Ave., Box 296
Malvern, PA 19355
(800) 345-1178, (215) 644-4412
Tools & restoration supplies

Fred Deming
13129 N. 22nd Ave.
Phoenix, AZ 85029
(602) 863-9233
New & used parts, restorations

Doug's Specialized Parts
7108 Crystal Valley Road
Little Rock, AR 72210

(501) 455-2562
Rebuilding, parts

Duke's Classic Mustang, Inc.
9313 Crowley Road
Ft. Worth, TX 76134
(817) 293-7114
Car dealer, parts, service, upholstery

Durbin's Mustangs
3864 S.R. 309
Gallon, OH 44833
(419) 468-7349
Car dealer, restoration, parts

East Coast Classic & Performance
1605 Gillespie
Fayetteville, NC 28306
(919) 483-3854
Parts & accessories

The Eastwood Company
580 Lancaster Ave., Box 296
Malvern, PA 19355
(800) 345-1178
Complete line of restoration tools

East Coast Antique Auto Parts
Box 330, Narrabeen
Sydney, NSW, Australia 2101
02/982-9305, 982-9335
Literature, parts, tires

Eastern Mustang Specialty
646 South Road
Poughkeepsie, NY 12601
(914) 462-6006, (914) 462-1094
Parts, accessories, literature

Ed Faxon Auto Literature
13955 E. Sixth St.
Building A, Suite 4B
Corona, CA 91719
(714) 735-6183
Literature

Florida Mustang Connection
6618 Highway 301 South
Riverview, FL 33569
(813) 677-3196
New & used parts, restoration

Ford Man's Mustangs & Automotive
10401 Royal Pine
Houston, TX 77093
(713) 697-6644
High-performance parts

The Ford Parts Warehouse
Courthouse Square
Liberty, KY 42539
(606) 787-5031
Parts

Free State Mustang
P.O. Box 10840
Baltimore, MD 21234
(301) 882-0131
New & used parts

Judd Freye
3240 Lakeshore Dr.
Muskegon, MI 49411
Accessories, used parts

Gelsi's Mustang World
3576 Northwest Blvd.
Vineland, NJ 08360
(609) 692-3178
New & used parts, restoration, car dealer

Glazier's Mustang Barn, Inc.
531 Wambold Road
Souderton, PA 18964
(800) 523-6708, (215) 723-9674
Complete restorations, parts

Granpa's Revenge Auto Rehabilitation
9228 Vista Dr.
Spring Valley, CA 92077
(619) 460-0961
Restoration, body work, painting

Herforth Motors
1686 Commonwealth Ave.
Boston, MA 02135
(617) 734-9057
Engine & transmission rebuilding, sandblasting, painting, welding

JBA Ford Performance
8280 Clairmont Mesa Blvd   127
San Diego, CA 92111
(619) 560-2030
High-performance parts

Jim's Mustangs
1399 Cuyamaca
El Cajon, CA 92020
(619) 562-0912
Rebuilt steering boxes

Just Suspension
P.O. Box 167
Towaco, NJ 07082
(201) 335-0547
Suspension parts

Jack Kaminski, Horsesense
1334 W. Piute
Phoenix, AZ 85027
(602) 869-7832
Car locator

Kanter Auto Products
76 Monroe St.
Boonton, NJ 07005
(800) 526-1096, (201) 334-9575
Parts

Ken's Falcon Parts
1799 E. Alosta
Glendora, CA 91740
(818) 963-5905
Parts

Lancaster Mustang
1042 Columbia Ave.
Lancaster, PA 17603
(717) 397-PONY
Parts & accessories

Larry's Thunderbird & Mustang Parts, Inc.
511 S. Raymond Ave.
Fullerton, CA 92631

(800) 854-0393, (714) 871-6432
Parts, upholstery

Last Chance Ford Parts
435 S. Main Street, Route 9
Forked River, NJ 08731
(609) 693-0333
Parts, engines, transmissions, accessories

Lee's Mustang
351 Buttonwood Lane
Cinnaminson, NJ 08077
(609) 829-6557
Fasteners, harnesses, problem solvers

Mackenzie Mustang Specialty
P.O. Box 857
Georgetown, MA 01833
Parts

Main Line Mustang
415 Maplewood Rd.
Wayne, PA 19087
(215) MUSTANG
Restoration

Mal's A Sales
4968 S. Pacheco Blvd
Martinex, CA 94553
(415) 228-8180
Parts

Mark's Mustangs
7202 W. Appleton Ave.
Milwaukee, WI 53216
(414) 464-5500
Car dealer, new & used parts

Maryland Mustang
1700 Millersville Rd.
Millersville, MD 21108
(301) 987-5353
Warehouse distributor

Marv's Fabulous Fords
10260 56th St.
Mira Loma, CA 91752
(714) 681-4545
Parts, restoration, car dealer

Mid-County Mustang, Inc.
P.O. Box 189
Eagle, PA 19480
(215) 458-8081
Restoration, parts, service

Mid-Michigan Mustang
7336 Division Ave. South
Grand Rapids, MI 49508
(616) 455-1340
New & used parts, restoration

Mostly Mustangs, Inc.
55 Alling St.
Hamden, CT 06517
(203) 562-8804, (203) 3062
Parts, restoration, car dealer

Mr. G's Mustang City
5613 Elliott Reeder Road
Ft. Worth, TX 76117
(817) 838-3131
Parts

Muscle Parts
P.O. Box 2579

Dearborn, MI 48123
(313) 291-3400
Parts

Muskegon Brake
848 E. Broadway
Muskegon, MI 49444
(616) 733-0874
Silicone brake fluid, springs

Mustang Of Chicago
1321 Irving Park Rd.
Bensenville, IL 60106
(312) 860-7077
Parts

Mustang Classics
2030 Vineyard Ave., Suite A & B
Escondido, CA 92025
(619) 741-3744
New & used parts (Shelby specialists)

Mustang Club of America
P.O. Box 447-MB
Lithonia, GA 30058-0447
National Mustang club

Mustang Collection
4912 SW 75th Ave.
Miami, FL 33155
(305) 264-4130
New & used parts

Mustang Connection of Kentucky, Inc.
4009 Crittenden Drive
Louisville, KY 40209
(502) 361-1447
Parts

Mustang Country
3210 Coffey Lane
Santa Rosa, CA 95401
(707) 526-4367
New & used parts

Mustang Custom
2030 Timber Lane
Dayton, OH 45414
Underbody parts, sheet metal

Mustang Emporium of San Diego
570 El Cajon Blvd
El Cajon, CA 92020
(619) 447-8908
Parts, accessories

Mustang Majic
5591 S. Rosemont Ave.
Tucson, AZ 85706
(602) 574-1268
Restoration, used parts

Mustang Man
Earl Parks
15514 E. Mustang Dr.
Fountain Hills, AZ 85268
Mail order parts

Mustang Mania
1522-1/2 S. 10th St.
Richmond, IN 47374
(317) 962-5248
Parts

Mustang Mart, Inc.
655 McGliney Lane
Campbell, CA 95008

(800) 538-7796, (408) 371-5771
Restoration supplies, new & used parts

Mustang Parts
Wayne T. Roth
9 Saddlemount Ave.
Warren, NJ 07060
(201) 755-2554
Convertible tops, parts, interior paint, upholstery

Mustang Parts Corral of Texas
P.O. Box 210524
Dallas, TX 75211
(214) 296-5130
Parts & accessories

Mustang Parts Of Oklahoma
6505 South Shields
Oklahoma City, OK 73149
(405) 631-1400
Full parts service

Mustang Ranch
36210 108 Ave. Court East
Eatonville, WA 98328
(206) 847-2623
New & used parts

Mustang Restorations, Inc.
1541 Brandy Parkway
Streamwood, IL 60103
(312) 830-8230
Parts, service, restoration

Mustang Stables
409 Bristol Pike
Croydon, PA 19020
Parts, painting, restoration

Mustang World
3802 Navy Blvd.
Pensacola, FL 32507
(904) 456-9333

Mustangs Only
1320 Oakland Rd.
San Jose, CA 95112
New & used parts

Mustangs Unlimited
185 Adams St.
Manchester, CT 06040
(800) 243-7278, (203) 647-1965
Parts, accessories, repair panels, restoration

National Parts Depot
1495-C Palma Dr.
Ventura, CA 93003
(800) 235-3445, (800) 342-3614
Parts, supplies, upholstery

National Parts Depot & Auto Craft
3101 SW 40th Blvd,
Gainesville, FL 32608
(800) 874-7595, (904) 378-2473
Parts, upholstery, restorations, car dealer

New England Mustang Supply
1830 Barnum Ave.
Bridgeport, CT 06610
(800) 242-2314, (203) 333-7454
Full parts service

Nu-Body Mustang, Inc.
73 Sandcreek Rd.
Albany, NY 12205
(518) 482-4184
Parts & accessories, body shop

Obsolete Ford Parts Co.
P.O. Box 787
Nashville, GA 31639
(912) 686-5101, (912) 686-2470
Parts

Ohio Mustang Connection
12981 National Rd. SW
Pataskala, OH 43062
(614) 927-6078
New & used parts, restoration

Old Timey Parts Co.
232 Lights Chapel Road
Breenbrier, TN 37073
(615) 643-7295
Restoration

Jim Osborn Reproductions, Inc.
101 Ridgecrest Dr.
Lawrenceville, GA 30245
(404) 962-7556
Literature, reproduction decals

The Paddock Inc.
P.O. Box 30
Knightstown, IN 46148
(317) 345-2131
Parts, supplies, body parts

Pennsylvania Mustang Parts
P.O. Box 660
Riegelsville, PA 18077
(610) 749-0411
1969-70 parts specialist

Performance Motorsports
P.O. Box 250
Valley Springs, SD 57068
(800) 843-9727
Stock & high performance parts

Pony and Corral
6102 Livingston Rd.
Oxon Hill, MD 20745
(301) 839-6666
New & used parts

Prestige Thunderbird, Inc.
10215 Greenleaf Ave.
Santa Fe Springs, CA 90670
(213) 944-6237
Accessories, car dealer, parts,
restoration

PRO Products
8460 Red Oak St., Suite B
Rancho Cucamonga, CA 91730-3815
(800) 826-0647, (800) 225-5985
Reproduction parts

Red's Headers
5832 Gibbons Dr.
Carmichael, CA 95608
Headers, mechanical parts, racing
parts

Rick's Antique Auto
2754 Roe Lane
Kansas City, KS 66103
(913) 722-5252
Hardware, parts, tools

SEMO Mustang
P.O. Box 78
Gordonville, MO 63752-0078
(314) 243-7664
Boss 302 reproductions, parts

Sacramento Mustang
5710 Auburn Blvd.
Sacramento, CA 95841
(800) 442-8333, (916) 334-0190
Parts & accessories

Salter's Antique Autos
235 Spruce St.
Milton, FL 32570
(904) 994-5189
Parts, restoration

Peter C. Sessler
235 E. 24th St.
New York, NY 10010
(212) 684-4210
High-performance parts

Carl Sprague Classic Custom
Autosound
300 E. Hermosa Dr.
Fullerton, CA 92635
(714) 738-8447
Radios

Speed-O-Motive
12061 Slauson Ave.
Santa Fe Springs, CA 90670
(213) 945-2758, (213) 945-3444
High-performance parts

Dan Stedem Ford, Inc.
S-3725 N. Buffalo Street
Orchard Park, NY 14127
(716) 662-4493
Parts

Stainless Steel Brakes Corp.
11470 Main Road
Clarence, NY 14031
(800) 448-7722, (716) 759-8666
S/S brake parts & conversion kits

Stillwell's Obsolete Car Parts
1617 Wedeking Avenue
Evansville, IN 47711
(812) 425-4794
Body parts, trim parts

Summitt's Used Cars & Parts
Rt. 3, Box 845, Wolfe Rd.
Kings Mountain, NC 28086
(704) 739-4909
Used parts

Sunwest Auto Specialties
510 General Hodges SE
Albuquerque, NM 87123
(505) 293-1751
Parts

Superior Mustang
1547 Palos Verdes Mall #114
Walnut Creek, CA 94596
(800) 282-4800, (415) 689-1770
Parts & accessories

J. Taylor
1114 Shamrock Dr.
Campbell, CA 95008
(408) 371-3508
Parts

Texas Mustang Parts
Rt 6, Box 996A
Waco, TX 76706
(817) 755-7948
Full parts service

T-Bird Parts Store, Inc.
12780 Currie Court
Livonia, MI 48150
(800) 521-6104, (313) 591-1956
Parts, tires

TMC Motorsprts
29171 W. Olympic Blvd.
Los Angeles, CA 90006
(213) 322-1131, (714) 966-1920
Suspension parts

Tubular Automotive Headers
248 Weymouth Street, Box 279
Rockland, MA 02370
(617) 878-9875
Headers

Valley Ford Parts Co.
11610 Vanowen St.
North Hollywood, CA 91605
(818) 982-5303
NOS & repro parts

The Vanishing Breed
2214 15th St.
Lubbock, TX 79401
(806) 763-3942
Car dealer, parts, parts repair

Vintage Ford Parts
3427 E. McDowell Road
Phoenix, AZ 85005
New & used parts

Vintage Mustangs
3440 Oakcliff Rd.
Doraville, GA 30340
(404) 455-9402

Virginia Mustang Supply, Inc.
Rt. 259 West, P.O. Box 487
Broadway, VA 22815
(703) 896-2695
Parts & accessories

Dan Williams 4-Speed Toploader
Transmissions
1210 NE 130th St., Dept. B
N. Miami, FL 33161
(305) 893-5123
Ford transmission repair, parts

METRIC CUSTOMARY-UNIT EQUIVALENTS

	Multiply:		by:		to get:		Multiply:		by:		to get:
LINEAR	inches	X	25.4	=	millimeters(mm)		X	0.03937	=	inches
	feet	X	0.3048	=	meters (m)		X	3.281	=	feet
	miles	X	1.6093	=	kilometers (km)		X	0.6214	=	miles
AREA	inches^2	X	645.16	=	millimeters^2(mm^2)		X	0.00155	=	inches^2
	inches^2	X	6.452	=	centimeters^2(m^2)		X	0.155	=	inches^2
VOLUME	inches^3	X	16387	=	millimeters^3(mm^3)		X	0.000061	=	inches^3
	quarts	X	0.94635	=	liters (l)		X	1.0567	=	quarts
	gallons	X	3.7854	=	liters (l)		X	0.2642	=	gallons
	feet^3	X	28.317	=	liters (l)		X	0.03531	=	feet^3
MASS	pounds (av)	X	0.4536	=	kilograms (kg)		X	2.2046	=	pounds (av)
FORCE	pounds—f(av)	X	4.448	=	newtons (N)		X	0.2248	=	pounds—f(av)
	kilograms—f	X	9.807	=	newtons (N)		X	0.10197	=	kilograms—f

METRIC CUSTOMARY-UNIT EQUIVALENTS (continued)

TEMPERATURE   Degrees Celsius (C) = 0.556 (F - 32)   Degrees Fahrenheit (F) = (1.8C) + 32

°F -40   0 32 40 80 98.6 120 160 200 212 240 280 320 °F

°C -40 -20   0   20   40   60   80 100 120 140 160 °C

PRESSURE OR STRESS	pounds/sq in.	X	6.895	=	kilopascals (kPa)	X 0.145	= pounds/sq in
	inches Hg (60F)	X	3.377	=	kilopascals (kPa)	X 0.2961	= inches Hg
POWER	horsepower	X	0.746	=	kilowatts (kW)	X 1.34	= horsepower
TORQUE	pound-inches	X	0.11298	=	newton-meters (N-m)	X 8.851	= pound-inches
	pound-feet	X	1.3558	=	newton-meters (N-m)	X 0.7376	= pound-feet
	pound-inches	X	0.0115	=	kilogram-meters (Kg-M)	X 87	= pound-inches
	pound-feet	X	0.138	=	kilogram-meters (Kg-M)	X 7.25	= pound-feet
VELOCITY	miles/hour	X	1.6093	=	kilometers/hour(km/h)	X 0.6214	= miles/hour

Index